Analytic Number Theory for Undergraduates

Monographs in Number Theory ISSN 1793-8341

Published

Vol. 1 Analytic Number Theory — An Introductory Course
by Paul T. Bateman & Harold G. Diamond

Vol. 2 Topics in Number Theory
by Minking Eie

Vol. 3 Analytic Number Theory for Undergraduates
by Heng Huat Chan

Monographs in

Volume 3

Analytic Number Theory for Undergraduates

Heng Huat Chan
National University of Singapore, Singapore

NEW JERSEY • LONDON • SINGAPORE • BEIJING • SHANGHAI • HONG KONG • TAIPEI • CHENNAI

Published by

World Scientific Publishing Co. Pte. Ltd.
5 Toh Tuck Link, Singapore 596224
USA office: 27 Warren Street, Suite 401-402, Hackensack, NJ 07601
UK office: 57 Shelton Street, Covent Garden, London WC2H 9HE

Acquisition Editor: Dr Phua Kok Khoo
Desk Editor: Kwong Lai Fun
In-house Series Editor: Tan Siew Lian
Designer: Mayrine Chung Mei Lan

British Library Cataloguing-in-Publication Data
A catalogue record for this book is available from the British Library.

ANALYTIC NUMBER THEORY FOR UNDERGRADUATES
Monographs in Number Theory — Vol. 3

ISBN-13 978-981-4271-35-6
ISBN-10 981-4271-35-7
ISBN-13 978-981-4271-36-3 (pbk)
ISBN-10 981-4271-36-5 (pbk)

Printed in Singapore.

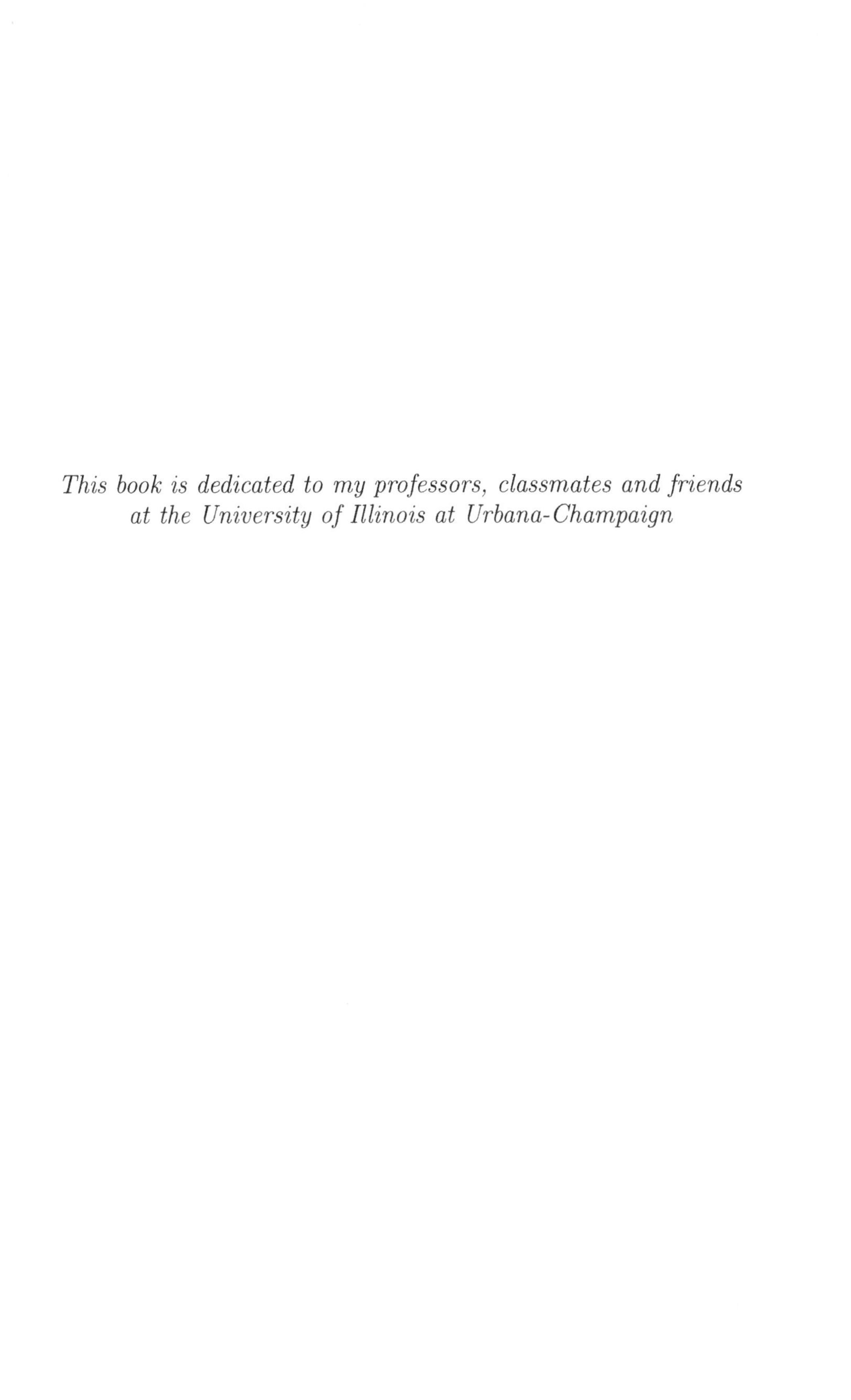

This book is dedicated to my professors, classmates and friends at the University of Illinois at Urbana-Champaign

Preface

I took several mathematics courses as a graduate student at University of Illinois at Urbana Champaign and one of the most interesting courses is A. Hildebrand's course titled "Analytic Number Theory 1". With Hildebrand's lecture notes and T.M. Apostol's book [2] as references, I designed an undergraduate level course on "Analytic Number Theory" and taught the course for several years between 2001 and 2007 at the National University of Singapore. The current book is a collection of notes arising from this course. The book is suitable for advanced undergraduate students who wish to understand number theory beyond elementary level.

There are seven chapters in this book. The materials in Chapter 1 are for readers who are not familiar with elementary number theory. The main focus of Chapter 2 is arithmetical functions while Chapter 3 discusses the averages of arithmetical functions. Elementary results related to the Prime Number Theorem are studied in Chapter 4 and the proof of the Prime Number Theorem is presented in Chapter 5.

Chapter 6 is devoted to the study of Dirichlet's series and the proof of Landau's Theorem. These materials are essential in the concluding chapter, whereby Dirichlet's Theorem on primes in arithmetic progression is proved.

I would like to thank A. Hildebrand for his permission to include part of his lecture notes in this book. I am also very grateful to B.C. Berndt for his advice on the preliminary version of this book.

I started writing this book in 2007. The final version was completed in Nagoya University when I was a Hitachi Fellow. I would like to thank the Hitachi Scholarship foundation for their generous support and my host at Nagoya University, Y. Tanigawa, for his warm hospitality.

Heng Huat Chan
24 December 2008

Contents

Chapter 1

The Fundamental Theorem of Arithmetic

1.1 Least Integer Axiom and Mathematical Induction

Let

$$\mathbf{Z} = \{0, \pm 1, \pm 2, \cdots\}$$

be the set of integers. The *Least Integer Axiom* (see [10]), also known as the *Well Ordering Principle*, states that there is a smallest integer in every *nonempty* subset of non-negative integers. It is useful in establishing the following result.

Theorem 1.1. *Let $S(1), S(2), \cdots, S(n), \cdots$ be statements, one for each integer $n \geq 1$. If some of these statements are false, then there is a first false statement.*

Proof. Set

$$T = \{k \in \mathbf{Z}^+ | S(k) \text{ is false}\}.$$

Since at least one statement is false, T is nonempty. By the Least Integer Axiom, there exists a smallest integer n in T. This implies that $S(n)$ is the first false statement. □

From Theorem 1.1, we deduce the *Principle of Mathematical Induction.*

Theorem 1.2. *Let $S(n)$ be statements, one for each $n \geq 1$. Suppose that the following conditions are satisfied by $S(n)$:*

(a) *The statement $S(1)$ is true.*
(b) *If $S(n)$ is true, then $S(n+1)$ is true.*

Then $S(n)$ is true for all integers $n \geq 1$.

Proof. Suppose that $S(n)$ is not true for all integers $n \geq 1$. Then for some positive integer $k \geq 1$, $S(k)$ is false. By Theorem 1.1, there is a first false statement, say $S(m)$. By the fact that $S(1)$ is true, we conclude that $m \neq 1$. Furthermore, by the minimality of m, we observe that $S(j)$ is true for $1 < j \leq m-1$. Now, by (b), $S(m-1)$ is true implies that $S(m)$ is true. This contradicts the assumption that $S(m)$ is false and we conclude that the statements $S(n)$ is true for all positive integers $n \geq 1$. □

We may replace 1 in Theorem 1.2 (a) by any integer m. In other words, we can modify Theorem 1.2 as

Theorem 1.3. *Let m be an integer. Let $S(n)$ be statements, one for each integer $n \geq m$. Suppose that the following two conditions are satisfied:*

(a) *The statement $S(m)$ is true.*
(b) *If $S(n)$ is true, then $S(n+1)$ is true.*

Then $S(n)$ is true for all integers $n \geq m$.

We end this section with another version of the Principle of Mathematical Induction. The proof of this version is similar to the proof of Theorem 1.2 and we leave it as an exercise for the readers.

Theorem 1.4. *Let m be an integer. Let $S(n)$ be statements, one for each integer $n \geq m$. Suppose that the following conditions are satisfied:*

(a) *$S(m)$ is true and*
(b) *if $S(k)$ is true for all $m \leq k \leq n$ then $S(n+1)$ is true.*

Then $S(n)$ is true for all integers $n \geq m$.

1.2 Division Algorithm

Theorem 1.5 (Division Algorithm). *Let a and b be integers such that $b > 0$. Then there exist unique integers q and r with*

$$a = bq + r, \quad \text{where } 0 \leq r < b.$$

Proof. Let

$$S = \{y | y = a - bx, \quad x \in \mathbf{Z} \text{ and } y \geq 0\}.$$

Note that since

$$a - b(-|a|) = a + b|a| \geq 0,$$

we find that

$$a + b|a| \in S,$$

and we conclude that S is nonempty. By the Least Integer Axiom, S contains a least non-negative integer, which we denote by r. We note that since $r \in S$,

$$r = a - bq,$$

for some integer q. We therefore conclude that

$$a = bq + r \quad \text{and} \quad r \geq 0.$$

We now show that $r < b$. Suppose $r \geq b$. Then

$$r - b \geq 0 \quad \text{and} \quad r - b = a - b(q + 1).$$

This implies that

$$r - b \in S.$$

By assumption, $b > 0$ and hence $r - b < r$. Hence, we have found a non-negative integer $r - b$ contained in S and smaller than r. This contradicts the minimality of r and we conclude that $r < b$.

Finally, we show that the integers q and r are unique. We suppose the contrary. Then there is a different representation of the form $a = bq' + r'$. This implies that

$$b(q' - q) = r - r' \tag{1.1}$$

and we conclude that $|r - r'|$ is a multiple of b. On the other hand, both $r, r' \in [0, b)$ and $|r - r'|$ can be a multiple of b only when $|r - r'| = 0$. In other words, $r = r'$ and by (1.1), $q = q'$. This contradicts the fact that the representations $a = bq' + r'$ $a = bq + r$ are different and therefore, the integers q and r must be unique.

□

Remark 1.1. The division algorithm Theorem 1.5 is also true for $b < 0$. We leave the details of the proof as an exercise.

When $r = 0$ in Theorem 1.5, we have $a = bq$ and we say that b divides a.

Definition 1.1. If an integer b divides a, we say that b is a *divisor* of a and that a is a *multiple* of b. The notation for b dividing a is given by

$$b|a.$$

Definition 1.2. We say that a positive integer is a *prime* if it has exactly two divisors, namely, 1 and itself.

We now state some elementary properties of divisibility.

Theorem 1.6. *Let a, b, d, m and n be nonzero integers. The following statements are true:*

(a) *For all nonzero integers k, $k|k$.*
(b) *If $d|n$ and $n|m$, then $d|m$.*
(c) *If $d|n$ and $d|m$, then $d|(an+bm)$.*
(d) *If $d|n$, then $ad|an$.*
(e) *If $ad|an$ and $a \neq 0$, then $d|n$.*
(f) *For all integers k, $1|k$.*
(g) *For all nonzero integers k, $k|0$.*
(h) *If $d|n$, then $|d| \leq |n|$.*
(i) *If $d|n$ and $n|d$, then $|d| = |n|$.*
(j) *If $d|n$, then $\left(\frac{n}{d}\right)|n$.*

Proof. We will prove (c) and leave the rest of the statements as exercises. Since $d|n$, we find that $n = ds$ for some integer s. Similarly, $d|m$ implies that $m = dt$ for some integer t. Now,

$$an + bm = ads + bdt = d(as + bt).$$

This shows that $d|(an+bm)$ for any integers a and b. □

In Theorem 1.6 (j), we see that if d is a divisor of n, then n/d is also a divisor of n. The divisor n/d is called the *conjugate divisor* of d.

1.3 Greatest common divisors

Definition 1.3. A *common divisor* of integers a and b is an integer c with $c|a$ and $c|b$.

Definition 1.4. A *greatest common divisor* of integers a and b is a number d with the following properties:

(a) The integer d is non-negative.
(b) The integer d is a common divisor of a and b.
(c) If e is any common divisor of a and b, then $e|d$.

Note that if d and d' are both greatest common divisors of a and b, then d is a common divisor of a and b and d' is a greatest common divisor, we note that $d'|d$ using Definition 1.4 (c). Similarly, since d' is a common divisor and d is a greatest common divisor, $d|d'$. By Theorem 1.6 (i), $|d| = |d'|$ and by Definition 1.4 (a), we deduce that $d = d'$. This shows that the greatest common divisor of a and b is unique.

The notation for the greatest common divisor of a and b is

$$(a, b).$$

Remark 1.2. When a and b are zeros, then $(0, 0) = 0$. If $a = 0$ and b is nonzero, then $(0, b) = b$.

We will next show that the greatest common divisor of two integers exists. By Remark 1.2, it suffices to consider the case when both a and b are nonzero.

Theorem 1.7. *Let a and b be nonzero integers. Then the smallest positive integer in the set*

$$P := \{sa + tb | s, t \in \mathbf{Z} \quad \textit{and} \quad sa + tb > 0\}$$

is (a, b).

Proof. If a is positive then $a \in P$ since

$$a = 1 \cdot a + 0 \cdot b.$$

Similarly, if b is positive, then $b \in P$. Suppose a and b are both negative. Then $0 \cdot a + (-1) \cdot b \in P$. Hence that P is nonempty. By the Least Integer Axiom, there is a smallest positive integer, say d, in P. Our aim is to show that

$$d = (a, b).$$

Since $d \in P$,

$$d = xa + yb \tag{1.2}$$

for some integers $x, y \in \mathbf{Z}$. We first show that d is a common divisor of a and b.

By Theorem 1.5, we may suppose

$$a = dq + r, \quad 0 \leq r < d.$$

Then

$$r = a - dq = a - (xaq + ybq) = a(1 - xq) + byq.$$

Therefore, $r \in P$ and it is smaller than d. But d is the smallest integer in P. Hence $r = 0$. In other words, $d|a$. By similar argument, with a replaced by b, we conclude that $d|b$. This shows that d is a common divisor of a and b.

Next, we observe that since $d \in P$, $d > 0$. Furthermore, if $c|a$ and $c|b$ then $a = cu$ and $b = cv$. This implies, by (1.2), that

$$d = xa + yb = c(ux + vy)$$

and hence, $c|d$. This shows that d satisfies the conditions in Definition 1.4 and we conclude that $d = (a, b)$. □

Identity (1.2) will be used frequently and we record it as follows:

Corollary 1.8. *Let a and b be integers. Then there exist integers x and y such that*

$$(a, b) = ax + by.$$

Definition 1.5. We say that two integers a and b are *relatively prime* if

$$(a, b) = 1.$$

Theorem 1.9. *Let a and b be nonzero integers. Then $(a, b) = 1$ if and only if $1 = ax + by$ for some integers x and y.*

Proof. Note that if $(a, b) = 1$, then by Corollary 1.8,

$$1 = ax + by$$

for some integers x and y.

Conversely, if

$$1 = ax + by,$$

then $(a, b)|a$ and $(a, b)|b$, and therefore $(a, b)|1$. This implies that $(a, b) = 1$. □

We now list down some basic properties of the greatest common divisor of two integers.

Theorem 1.10. *Let a, b and c be nonzero integers. Then*

(a) $(a, b) = (b, a)$
(b) $(a, (b, c)) = ((a, b), c)$ *and*
(c) $(ac, bc) = |c|(a, b)$.

Proof. We will prove only (c) and leave the proofs of the other statements as exercises. Let $d = (ac, bc)$ and $d' = |c|(a, b)$. By Corollary 1.8,

$$d = acx + bcy$$

for some integers x and y. Hence,

$$d = \frac{c}{|c|}\left(a \cdot |c| \cdot x + b \cdot |c| \cdot y\right). \tag{1.3}$$

Now, $d' = |c|(a, b)$ and since $(a, b)|a$ and $(a, b)|b$, we find that d' is a common divisor of $a \cdot |c|$ and $b \cdot |c|$ and therefore, by (1.3), $d'|d$.

Next, since $d'/|c| = (a, b)$, by Corollary 1.8,

$$\frac{d'}{|c|} = au + bv$$

for some integers u and v. This implies that

$$d' = a \cdot |c| \cdot u + b \cdot |c| \cdot v = \frac{|c|}{c}\left(acu + bcv\right).$$

But d is a common divisor of ac and bc and hence $d|d'$. Since $d'|d$ and $d|d'$, we conclude by Theorem 1.6 (i) that $|d| = |d'|$. Since both d and d' are positive, we deduce that $d = d'$. □

1.4 The Euclidean Algorithm

In this section, we prove a result that allows us to compute the greatest common divisor of two integers. First, we need a lemma.

Lemma 1.11. *Let a, b, q, and r be integers such that*

$$a = bq + r,$$

then

$$(a, b) = (b, r).$$

Proof. Let $d = (a, b)$ and $d' = (b, r)$. Note that since $d|a$ and $d|b$, we find that $d|(a - bq)$ by Theorem 1.6 (c). Hence, $d|r$ and d is a common divisor of b and r. By Definition 1.4 (c), $d|d'$ since $d' = (b, r)$. Similarly, $d'|b$ and $d'|r$ implies that $d'|(bq + r)$ by Theorem 1.6 (c) and consequently, $d'|a$. By Definition 1.4 (c), $d'|d$ since $d = (a, b)$. Therefore, by Theorem 1.6 (i), $d = d'$. □

Theorem 1.12 (The Euclidean Algorithm). *Given positive integers a and b, where $b \nmid a$. Let $r_0 = a, r_1 = b$, and apply the division algorithm repeatedly to obtain a set of remainders $r_2, r_3, ..., r_n, r_{n+1}$ defined successively by the relations*

$$\begin{array}{lll} r_0 & = r_1 q_1 + r_2 & 0 < r_2 < r_1 \\ r_1 & = r_2 q_2 + r_3 & 0 < r_3 < r_2 \\ & \vdots & \\ r_{n-2} & = r_{n-1} q_{n-1} + r_n & 0 < r_n < r_{n-1} \\ r_{n-1} & = r_n q_n + r_{n+1} & r_{n+1} = 0. \end{array}$$

Then r_n, the last nonzero remainder in this process is (a, b), the greatest common divisor of a and b.

Proof. There is a stage at which $r_{n+1} = 0$ because the r_i are decreasing and non-negative. Next, applying Lemma 1.11, we find that

$$(a, b) = (r_0, r_1) = (r_1, r_2) = \cdots = (r_n, r_{n+1}) = (r_n, 0) = r_n.$$

This completes the proof of Theorem 1.12. □

Example 1.1. Find $(196884, 576)$.

Solution.

$$\begin{aligned} 196884 &= 341 \cdot 576 + 468 \\ 576 &= 1 \cdot 468 + 108 \\ 468 &= 4 \cdot 108 + 36 \\ 108 &= 3 \cdot 36 + 0. \end{aligned}$$

Therefore $(196884, 576) = 36$.

1.5 Congruences

We say that a is congruent to b modulo n when $n|(a - b)$. The notation is

$$a \equiv b \pmod{n}.$$

Theorem 1.13 (Basic Properties of Congruences). *Let a, b, c, d, n be integers with $n > 0$. Then*

(a) *For all integers k, $k \equiv k \pmod{n}$.*
(b) *If $a \equiv b \pmod{n}$ then $b \equiv a \pmod{n}$.*
(c) *If $a \equiv b \pmod{n}$ and $b \equiv c \pmod{n}$ then $a \equiv c \pmod{n}$.*
(d) *If $a \equiv b \pmod{n}$ and $c \equiv d \pmod{n}$ then $a + c \equiv b + d \pmod{n}$ and $ac \equiv bd \pmod{n}$.*

The proof of Theorem 1.13 is straightforward and we leave it as an exercise.

We know that if $c \neq 0$ then $ca = cb$ implies that $a = b$. This is known as the law of cancellation for equality. The law is not true in general if we replace "=" by $\equiv$. For example, $15 \equiv 3 \pmod{12}$ but $5 \not\equiv 1 \pmod{12}$. The next result shows that the law of cancellation holds if we impose a condition on the integer c.

Theorem 1.14. *Let a, b, c and n be integers. If $ca \equiv cb \pmod{n}$ and $(c, n) = 1$, then $a \equiv b \pmod{n}$.*

Proof. Recall from Corollary 1.8 that if $(c, n) = 1$ then there exist integers x and y such that $cx + ny = 1$. Multiplying a and b yields

$$acx + any = a$$

and

$$bcx + bny = b,$$

respectively. Since $ac \equiv bc \pmod{n}$, we conclude that $a - b \equiv (ac - bc)x \equiv 0 \pmod{n}$ and hence,

$$a \equiv b \pmod{n}. \qquad \square$$

Theorem 1.14 can be used to prove the following result of Euclid.

Corollary 1.15 (Euclid's Lemma). *Let a and b be integers and p be a prime. If $p|(ab)$, then $p|a$ or $p|b$.*

Proof. For any integer n, $(n, p) = 1$ or p since p has only two divisors. Suppose $p \nmid a$. Then (p, a)=1. By Theorem 1.14, the relation

$$ab \equiv 0 \pmod{p}$$

then implies that

$$b \equiv 0 \pmod{p}. \qquad \square$$

By induction, we have the following:

Corollary 1.16. *Let $a_1, a_2, \cdots, a_m$ be integers and let p be a prime. If $p|(a_1 a_2 \cdots a_m)$ then $p|a_k$ for some k.*

1.6 Fundamental Theorem of Arithmetic

Theorem 1.17 (Fundamental Theorem of Arithmetic). *Every positive integer $n > 1$ can be expressed as a product of primes; this representation is unique apart from the order in which the factors occur.*

Proof. We first show that n can be expressed as a prime or a product of primes. We use induction on n. The statement is clearly true for $n = 2$ since 2 is a prime. Suppose m is a prime or a product of primes for $2 \leq m \leq n-1$. If n is a prime then we are done. Suppose n is composite then $n = ab$, where $1 < a, b < n$. By induction each of the a and b is either a prime or a product of primes. Hence, $n = ab$ is a product of primes. By Theorem 1.4, every positive integer $n > 1$ is a prime or a product of primes.

To prove uniqueness, we use induction on n again. If $n = 2$ then the representation of n as a product of primes is clearly unique. Assume, then that it is true for all integers greater than 1 and less than n. We shall prove that it is also true for n. If n is prime, then there is nothing to prove. Assume, then, that n is composite and that n has two factorizations, say,

$$n = p_1 p_2 \cdots p_s = q_1 q_2 \cdots q_t. \tag{1.4}$$

Since p_1 divides the product $q_1 q_2 \cdots q_t$, it must divide at least one factor by Corollary 1.16. Relabel $q_1, q_2, ..., q_t$ so that $p_1|q_1$. Then $p_1 = q_1$ since both p_1 and q_1 are primes. In (1.4), we may cancel p_1 on both sides to obtain

$$n/p_1 = p_2 \cdots p_s = q_2 \cdots q_t.$$

Now the induction hypothesis implies that the two factorizations of n/p_1 must be the same, apart from the order of the factors. Therefore, $s = t$ and the factorizations in (1.4) are also identical, apart from order. This completes the proof. □

In subsequent chapters, whenever we write

$$n = p_1^{\alpha_1} p_2^{\alpha_2} \cdots p_r^{\alpha_r},$$

we mean that $p_1^{\alpha_1}p_2^{\alpha_2}\cdots p_r^{\alpha_r}$ is the prime power decomposition of n that is unique up to rearrangement of factors. When we write

$$n = \prod_{k=1}^{r} p_k^{\alpha_k}$$

we mean that $\alpha_j \neq 0, 1 \leq j \leq r$. If we write

$$n = \prod_{p} p^{\alpha_p},$$

then we understand that only finitely many α_p's are nonzero.

1.7 Exercises

1. Complete the proofs of Theorems 1.6, 1.10 and 1.13.
2. Let a, b and c be nonzero integers. If $(a,b) = (a,c) = 1$ then $(a,bc) = 1$.
3. Let a, b, x, y be nonzero integers and let n be a positive integer. Show that if $(a,b) = 1$ and $ab = c^n$, then $a = x^n$ and $b = y^n$ for some x and y.
4. Prove that if n is a positive integer and $2^n + 1$ is a prime, then n is a power of 2.
5. Let a, m, and n be positive integers. Show that $(a^m - 1, a^n - 1) = a^{(m,n)} - 1$.
6. The least common multiple of two nonzero integers a, b is defined as

$$[ab] = |ab|/(a,b).$$

   Show that

$$(a,[b,c]) = [(a,b),(a,c)].$$

7. Let a, b be integers. Show that if $(a,b) = 1$, then $(a + b, a^2 - ab + b^2)$ is either 1 or 3.
8. Let n be a positive integer. Show that if $2^n - 1$ is prime, then n is a prime.
9. Let m and n be positive integers and $m \neq n$. Find (A_m, A_n) where $A_m = a^{2^m} + 1$.

10. Let n and k be positive integers. Show that if $(n-1)^2|(n^k-1)$, then $(n-1)|k$. Hence, or otherwise, show that if p is a prime, then the following statements are equivalent:

    (a) $(p-1)!+1$ is a power of p.

    (b) $p = 2, 3$ and 5.

Chapter 2

Arithmetical Functions and Dirichlet Multiplication

2.1 The Möbius function

Definition 2.1. A real or complex-valued function defined on the set of positive integers is called an *arithmetical function.*

Arithmetical functions play an important role in the study of numbers. Let us now introduce one of the most important arithmetical functions, namely, the Möbius function $\mu(n)$.

Definition 2.2. Let $\mu(1) = 1$. If $n = p_1^{\alpha_1} \cdots p_k^{\alpha_k}$, then define

$$\mu(n) = \begin{cases} (-1)^k & \text{if } \alpha_1 = \alpha_2 = \cdots = \alpha_k = 1, \\ 0 & \text{otherwise.} \end{cases}$$

The function $\mu(n)$ is known as the Möbius function.

We observe that by the above definition, the Möbius function $\mu(n)$ is identically zero if and only if n has a square factor greater than 1.

Definition 2.3. We write

$$\sum_{d|n} f(d)$$

to denote the sum of the values of f over divisors d of n.

Remark 2.1. Suppose $d_1, d_2, \cdots, d_k$ are the divisors of n, then

$$\sum_{d|n} f(d) = f(d_1) + f(d_2) + \cdots + f(d_k).$$

Now, each d_j can be written in the form n/d'_j, where d'_j is the conjugate divisor of d_j. Hence,

$$\sum_{d|n} f(d) = f\left(\frac{n}{d'_1}\right) + f\left(\frac{n}{d'_2}\right) + \cdots + f\left(\frac{n}{d'_k}\right) = \sum_{d|n} f\left(\frac{n}{d}\right).$$

Therefore

$$\sum_{d|n} f(d) = \sum_{d|n} f\left(\frac{n}{d}\right). \tag{2.1}$$

Theorem 2.1. *Let n be any positive integer and $[x]$ denote the integer part of a real number x. We have*

$$\sum_{d|n} \mu(d) = \left[\frac{1}{n}\right] = \begin{cases} 1 & \text{if } n = 1, \\ 0 & \text{if } n > 1. \end{cases}$$

Proof. The formula is true for $n = 1$. Assume that $n > 1$ and

$$n = p_1^{\alpha_1} \cdots p_k^{\alpha_k}.$$

In the sum

$$\sum_{d|n} \mu(d),$$

the terms are nonzero when $d = 1$ or when the divisors of n are products of distinct primes. Hence, we find that

$$\begin{aligned} \sum_{d|n} \mu(d) &= \sum_{d|(p_1 p_2 \cdots p_k)} \mu(d) \\ &= \mu(1) + \mu(p_1) + \cdots + \mu(p_k) + \mu(p_1 p_2) + \cdots + \mu(p_{k-1} p_k) \\ &\quad + \cdots + \mu(p_1 \cdots p_k). \end{aligned} \tag{2.2}$$

Since there are $\binom{k}{j}$ ways for choosing j primes from a set of k primes, we conclude from (2.2) that

$$\sum_{d|n} \mu(d) = 1 + \binom{k}{1}(-1) + \binom{k}{2}(-1)^2 + \cdots + \binom{k}{k}(-1)^k = (1-1)^k = 0.$$

□

Definition 2.4. For any positive integer $n \geq 1$, we define

$$I(n) = \left[\frac{1}{n}\right].$$

With Definition 2.4, we may rewrite Theorem 2.1 as

$$\sum_{d|n} \mu(d) = I(n). \tag{2.3}$$

2.2 The Euler totient function

Definition 2.5. The Euler totient $\varphi(n)$ is defined to be the number of positive integers not exceeding n which are relatively prime (see Definition 1.5) to n.

It is sometimes convenient to write $\varphi(n)$ as

$$\varphi(n) = \sum_{\substack{k=1 \\ (k,n)=1}}^{n} 1. \tag{2.4}$$

The first important result for $\varphi(n)$ is the following theorem:

Theorem 2.2. *Let n be any positive integer. Then*

$$\sum_{d|n} \varphi(d) = n.$$

Proof. Let

$$S = \{k \in \mathbf{Z} \,|\, 1 \leq k \leq n\}$$

and

$$A(d) = \{k \in \mathbf{Z} \,|\, (k,n) = d, 1 \leq k \leq n\}.$$

Since every integer $k \leq n$ has a unique (k,n), we conclude that S is a disjoint union of $A(d)$ and we deduce that

$$\sum_{d|n} f(d) = n, \tag{2.5}$$

where $f(d)$ be the number of elements in $A(d)$.

Let

$$B(d) = \{1 \leq q \leq n/d \,|\, (q, n/d) = 1\}.$$

We claim that there is a one to one correspondence between the elements in $A(d)$ and $B(d)$. If $1 \leq k \leq n$, then the element $k/d \in B(d)$ since

$$(k/d, n/d) = 1,$$

by Theorem 1.10 (c).

Conversely, given $q \in B(d)$, let $q = k/d$. Then $k \in A(d)$.

Now, the number of elements in $B(d)$ is $\varphi(n/d)$. Hence, by (2.5) and the fact that

$$|A(d)| = |B(d)|,$$

where $|U|$ is the number of elements of the set U, we find that $f(d) = \varphi(n/d)$. Therefore,

$$\sum_{d|n} \varphi\left(\frac{n}{d}\right) = n.$$

But by (2.1), this is equivalent to

$$\sum_{d|n} \varphi(d) = n.$$

□

Next, we establish a relation between $\varphi(n)$ and $\mu(n)$.

Theorem 2.3. *Let n be any positive integer. Then*

$$\varphi(n) = \sum_{d|n} \mu(d)\frac{n}{d}.$$

Proof. We first observe that

$$I(j) = \begin{cases} 1 & \text{if } j = 1, \\ 0 & \text{if } j > 1. \end{cases}$$

Therefore, if $g(k)$ is an arithmetical function, then

$$\sum_{\substack{k=1 \\ (k,n)=1}}^{n} g(k) = \sum_{k=1}^{n} g(k)I((k,n)).$$

Setting $g(k) = 1$, we find that

$$\varphi(n) = \sum_{\substack{k=1 \\ (k,n)=1}}^{n} 1 = \sum_{k=1}^{n} I((k,n)).$$

By (2.3), we deduce that

$$\varphi(n) = \sum_{k=1}^{n} I((k,n)) = \sum_{k=1}^{n} \sum_{d|(k,n)} \mu(d) = \sum_{k=1}^{n} \sum_{\substack{d|k \\ d|n}} \mu(d)$$

$$= \sum_{d|n} \mu(d) \sum_{q=1}^{n/d} 1 = \sum_{d|n} \mu(d)\frac{n}{d}.$$

This completes the proof of the theorem. □

Given a positive integer n, we can always compute $\varphi(n)$ directly by counting those integers $k \leq n$ that are relatively prime to n. However, this direct method is very tedious except for the cases when n is a prime or a prime power.

If p is a prime, then all positive integers less than p are relatively prime and we deduce that

$$\varphi(p) = p - 1. \tag{2.6}$$

If $n = p^\alpha$ then the only positive integers less than p^α that *are not relatively prime* to p^α are multiples of p. There are exactly $p^{\alpha-1}$ such integers and hence,

$$\varphi(p^\alpha) = p^\alpha - p^{\alpha-1}. \tag{2.7}$$

When n is not a prime or a prime power, the computation of $\varphi(n)$ is given by the following theorem. Note that the formula for computing $\varphi(n)$ is only useful when we know the prime factorization of n.

Theorem 2.4. *Let n be any positive integer with prime factorization*

$$n = \prod_{j=1}^{k} p_j^{\alpha_j}.$$

Then

$$\varphi(n) = n \prod_{p|n} \left(1 - \frac{1}{p}\right).$$

Proof. By Theorem 2.3, we find that

$$\varphi(n) = n \sum_{d|n} \frac{\mu(d)}{d} = n \sum_{d|p_1^{\alpha_1}\cdots p_k^{\alpha_k}} \frac{\mu(d)}{d} = n \sum_{d|p_1\cdots p_k} \frac{\mu(d)}{d}. \tag{2.8}$$

The last equality of (2.8) follows from the fact that $\mu(d)$ is nonzero when d is 1 or a squarefree divisor of n. Hence,

$$\begin{aligned}\varphi(n) &= n\left(1 + \sum_{p_i} \frac{\mu(p_i)}{p_i} + \sum_{p_i, p_j} \frac{\mu(p_i p_j)}{p_i p_j} + \cdots + \frac{\mu(p_1 \cdots p_k)}{p_1 \cdots p_k}\right) \\ &= n\left(1 - \sum_{p_i} \frac{1}{p_i} + \sum_{p_i, p_j} \frac{1}{p_i p_j} - \cdots + \frac{(-1)^k}{p_1 \cdots p_k}\right).\end{aligned}$$

Next, recall that if $a_1, a_2, \cdots, a_k$ are distinct, then

$$\begin{aligned}(u-a_1)&(u-a_2)\cdots(u-a_k)\\ &= u^k - \sum_i a_i u^{k-1} + \sum_{i,j} a_i a_j u^{k-2} + \cdots + (-1)^k a_1 a_2 \cdots a_k.\end{aligned}$$

With $u = 1$ and $a_j = 1/p_j$, we conclude that

$$\begin{aligned}\varphi(n) &= n\left(1 - \sum_{p_i} \frac{1}{p_i} + \sum_{p_i,p_j} \frac{1}{p_i p_j} \cdots + \frac{(-1)^k}{p_1 \cdots p_k}\right)\\ &= n\left(1-\frac{1}{p_1}\right)\left(1-\frac{1}{p_2}\right)\cdots\left(1-\frac{1}{p_k}\right)\\ &= n\prod_{j=1}^{k}\left(1-\frac{1}{p_j}\right).\end{aligned}$$

Note that the product

$$\prod_{j=1}^{k}\left(1-\frac{1}{p_j}\right) = \prod_{p|n}\left(1-\frac{1}{p}\right),$$

and we complete the proof of the theorem. □

Several properties of $\varphi(n)$ follow immediately from Theorem 2.4.

Corollary 2.5.

(a) *Let p be a prime number and α be any positive integer. Then*

$$\varphi(p^\alpha) = p^\alpha - p^{\alpha-1}.$$

(b) *If m and n are positive integers such that $d = (m, n)$, then $\varphi(mn) = \varphi(m)\varphi(n)\left(\frac{d}{\varphi(d)}\right)$.*

(c) *If m and n are relatively prime, then $\varphi(mn) = \varphi(m)\varphi(n)$.*

Proof.

(a) Note that the formula is (2.7) which we have computed earlier. This identity follows immediately from Theorem 2.4 by setting $n = p^\alpha$.

(b) Let m and n be positive integers and $d = (m, n)$. Write $m = m'd_1$ and $n = n'd_2$, where d_1 and d_2 are products of primes that divide d. By the choice of d_1,

$$(m', d_1) = 1. \tag{2.9}$$

If $(m', n') > 1$, then a prime p dividing (m', n') would divide $(m, n) = d$. This implies that $p|d_1$. But $p \nmid m'$ and hence, $(m', d_1) \neq 1$, which contradicts (2.9). Hence, we conclude that $(m', n') = 1$.
By the choices of d_1 and d_2, we deduce that

$$\prod_{p|d}\left(1-\frac{1}{p}\right) = \prod_{p|d_1}\left(1-\frac{1}{p}\right) = \prod_{p|d_2}\left(1-\frac{1}{p}\right) = \prod_{p|d_1d_2}\left(1-\frac{1}{p}\right). \tag{2.10}$$

This implies that

$$\begin{aligned}\varphi(mn) &= mn\prod_{p|(mn)}\left(1-\frac{1}{p}\right)\\ &= mn\prod_{p|(m'n'd_1d_2)}\left(1-\frac{1}{p}\right)\\ &= mn\prod_{p|m'}\left(1-\frac{1}{p}\right)\prod_{p|n'}\left(1-\frac{1}{p}\right)\prod_{p|d}\left(1-\frac{1}{p}\right),\end{aligned}$$

by (2.10). Hence,

$$\begin{aligned}\varphi(mn) &= mn\prod_{p|m'}\left(1-\frac{1}{p}\right)\prod_{p|n'}\left(1-\frac{1}{p}\right)\prod_{p|d}\left(1-\frac{1}{p}\right)\frac{\prod_{p|d}\left(1-\frac{1}{p}\right)}{\prod_{p|d}\left(1-\frac{1}{p}\right)}\\ &= mn\frac{\prod_{p|m'}\left(1-\frac{1}{p}\right)\prod_{p|n'}\left(1-\frac{1}{p}\right)\prod_{p|d_1}\left(1-\frac{1}{p}\right)\prod_{p|d_2}\left(1-\frac{1}{p}\right)}{\prod_{p|d}\left(1-\frac{1}{p}\right)}\\ &= mn\frac{\prod_{p|m}\left(1-\frac{1}{p}\right)\prod_{p|n}\left(1-\frac{1}{p}\right)}{\prod_{p|d}\left(1-\frac{1}{p}\right)}\\ &= \varphi(m)\varphi(n)\left(\frac{d}{\varphi(d)}\right),\end{aligned}$$

by (2.10).

(c) Set $d = 1$ in (b). □

Definition 2.6. An arithmetical function f is said to be *multiplicative* if

$$f(1) = 1$$

and

$$f(mn) = f(m)f(n) \quad \text{whenever} \quad (m, n) = 1.$$

Corollary 2.5 (c) shows that the Euler φ function is multiplicative. Other examples of multiplicative functions are the Möbius function $\mu(n)$ and $I(n)$. We leave these facts as exercises for the readers.

Suppose $n > 1$ is an integer written in the form

$$n = \prod_{i=1}^{k} p_i^{\alpha_i}$$

and if f is multiplicative, then

$$f\left(\prod_{i=1}^{k} p_i^{\alpha_i}\right) = \prod_{i=1}^{k} f(p_i^{\alpha_i}).$$

This shows that if f is multiplicative, then its value at any positive integer n is determined by its values at prime powers.

2.3 Dirichlet product of arithmetical functions and multiplicative functions

Definition 2.7. Let f and g be two arithmetical functions. We define the *Dirichlet product* of f and g, denoted by $f * g$, as

$$(f * g)(n) = \sum_{d|n} f(d) g\left(\frac{n}{d}\right).$$

We will often use $f * g$ to represent the function $(f * g)(n)$, suppressing the argument n.

Definition 2.8. Let n be any positive integer. The arithmetical function $N(n)$ is defined by

$$N(n) = n.$$

Using the above notation, Definitions 2.7 and 2.8, Theorem 2.3 can simply be written as

$$\varphi = \mu * N.$$

Theorem 2.6. *The Dirichlet product is commutative and associative, that is, for any arithmetical functions f, g, k, we have*

$$f * g = g * f$$

and

$$(f * g) * k = f * (g * k).$$

Proof. The Dirichlet product of f and g is given by

$$(f * g)(n) = \sum_{d|n} f(d) g\left(\frac{n}{d}\right).$$

Let $d_1 = n/d$ be the conjugate divisor of d. As d runs through all divisors of n, so does d_1. By (2.1),

$$(f * g)(n) = \sum_{d_1|n} f\left(\frac{n}{d_1}\right) g(d_1) = (g * f)(n).$$

To prove the associativity property, let $A = g * k$. Then

$$\begin{aligned}(f * A)(n) &= \sum_{a|n} f(a) A\left(\frac{n}{a}\right) \\ &= \sum_{a \cdot d = n} f(a) \sum_{b \cdot c = d} g(b) k(c) \\ &= \sum_{a \cdot b \cdot c = n} f(a) g(b) k(c).\end{aligned}$$

Similarly, if we set $B = (f * g)$, then

$$\begin{aligned}(B * k)(n) &= \sum_{d \cdot c = n} B(d) k(c) \\ &= \sum_{d \cdot c = n} \sum_{a \cdot b = d} f(a) g(b) k(c) \\ &= \sum_{a \cdot b \cdot c = n} f(a) g(b) k(c).\end{aligned}$$

Therefore,

$$(f * (g * k))(n) = ((f * g) * k)(n).$$

□

Theorem 2.7. *Let $I(n)$ be given by Definition 2.4. The function I is the identity function for $*$, that is, $I * f = f * I = f$ for every arithmetical function f.*

Proof. By the definition of I, we find that

$$(I * f)(n) = \sum_{d|n} I(d) f\left(\frac{n}{d}\right) = f(n).$$

By the commutative law in Theorem 2.6, we conclude that

$$f * I = f.$$

□

2.4 Dirichlet inverses and the Möbius inversion formula

Theorem 2.8. *Let f be an arithmetical function. If $f(1) \neq 0$, then there is a unique function g such that*

$$f * g = I. \tag{2.11}$$

Proof. We show by induction on m that (2.11) has a unique solution $g(m)$. In order for (2.11) to hold, the function $g(n)$ must satisfy

$$f(1)g(1) = 1.$$

Since $f(1) \neq 0$, we find that

$$g(1) = \frac{1}{f(1)}$$

and $g(1)$ is uniquely determined. Suppose $m > 1$ and assume that the values of $g(k)$ have been determined for $1 \leq k \leq m-1$. From (2.11), we find that

$$f(1)g(m) + \sum_{\substack{d|m \\ d>1}} f(d)g\left(\frac{m}{d}\right) = 0.$$

Therefore,

$$g(m) = \frac{1}{f(1)}\left(-\sum_{\substack{d|m \\ d>1}} f(d)g\left(\frac{m}{d}\right)\right)$$

and $g(m)$ is uniquely determined. By mathematical induction, there is a unique function $g(n)$ such that

$$f * g = I.$$

□

Definition 2.9. Given an arithmetical function f such that $f(1) \neq 0$. The unique function g such that $f * g = I$ is called the *Dirichlet inverse* of f. The notation for the Dirichlet inverse of f is f^{-1}.

We next introduce another arithmetical function.

Definition 2.10. Let n be any positive integer. The arithmetical function u is defined by

$$u(n) = 1.$$

With the above notation, Theorem 2.1 can be stated as

$$\mu * u = I.$$

Therefore, we have $u = \mu^{-1}$ and $\mu = u^{-1}$. In other words, we have

Theorem 2.9. *The Dirichlet inverse of* u *is* μ.

Theorem 2.10 (Möbius inversion formula). *If* $f = g * u$, *then* $g = f * \mu$. *Conversely,* $g = f * \mu$ *implies that* $f = g * u$.

Proof. Suppose $f = g * u$. Then by Theorem 2.9,

$$f * \mu = (g * u) * \mu = g * (u * \mu) = g * I = g.$$

Conversely, if $g = f * \mu$ then

$$g * u = (f * \mu) * u = f * (\mu * u) = f * I = f. \qquad \square$$

Theorems 2.2 and 2.3 are special cases of the Möbius inversion formula. In other words,

$$N = \varphi * u \quad \text{if and only if} \quad N * \mu = \varphi.$$

2.5 Multiplicative functions and Dirichlet products

From the construction of f^{-1} in Theorem 2.8, it is not clear that the Dirichlet inverse of a multiplicative function f is multiplicative. In the next few results, we will show that this is indeed the case.

Theorem 2.11. *If* f *and* g *are multiplicative functions, then so is their Dirichlet product* $f * g$.

Proof. Let $h = f * g$. Note that

$$h(1) = f(1)g(1) = 1.$$

Next, consider the expression

$$h(mn) = \sum_{c|mn} f(c) g\left(\frac{mn}{c}\right).$$

Given that $(m, n) = 1$, we can write $c = ab$, where $a|m$ and $b|n$. Therefore, we deduce that

$$\begin{aligned} h(mn) &= \sum_{a|m} \sum_{b|n} f(ab) g\left(\frac{m}{a}\frac{n}{b}\right) \\ &= \sum_{a|m} \sum_{b|n} f(a) f(b) g\left(\frac{m}{a}\right) g\left(\frac{n}{b}\right), \end{aligned}$$

since $(m/a, n/b) = 1$ and both f and g are multiplicative. This implies that

$$h(mn) = \sum_{a|m} f(a)g\left(\frac{m}{a}\right) \sum_{b|n} f(b)g\left(\frac{n}{b}\right)$$
$$= h(m)h(n).$$

□

Theorem 2.12. *If both g and $f * g$ are multiplicative, then f is also multiplicative.*

Proof. We prove the theorem by contradiction. Suppose f is not multiplicative. Let

$$h = f * g.$$

Since f is not multiplicative, there exist two relatively prime integers m and n such that

$$f(mn) \neq f(m)f(n).$$

We choose mn as small as possible. If $mn = 1$, then

$$f(1) \neq f(1)f(1)$$

and $f(1) \neq 1$. Since $h(1) = f(1)g(1) = f(1) \neq 1$, we conclude that h is not multiplicative, which leads to a contradiction. Hence, $mn \neq 1$.

If $mn > 1$, then

$$f(ab) = f(a)f(b)$$

for all $1 \leq ab < mn$ and $(a, b) = 1$. Now,

$$h(mn) = f(mn)g(1) + \sum_{\substack{a|m \\ b|n \\ ab<mn}} f(ab)g\left(\frac{mn}{ab}\right)$$
$$= f(mn) + \sum_{\substack{a|m \\ b|n \\ ab<mn}} f(a)f(b)g\left(\frac{m}{a}\right)g\left(\frac{n}{b}\right)$$
$$= f(mn) - f(m)f(n) + h(m)h(n).$$

In other words,

$$h(mn) - h(m)h(n) = f(mn) - f(m)f(n).$$

Since $f(mn) \neq f(m)f(n)$, we find that $h(mn) \neq h(m)h(n)$. This implies that h is not multiplicative, which contradicts our assumption that h is multiplicative. □

Theorem 2.13. *If g is multiplicative, then the Dirichlet inverse g^{-1} is also multiplicative.*

Proof. The functions g and $g * g^{-1} = I$ are multiplicative. By Theorem 2.12, g^{-1} is multiplicative. □

Remark 2.2. We have shown here that the set of multiplicative arithmetical functions forms an abelian group under the Dirichlet product $*$, with identity I.

2.6 Exercises

1. (von Mangoldt's function $\Lambda(n)$) For every integer $n \geq 1$, we define

$$\Lambda(n) = \begin{cases} \ln p & \text{if } n = p^m \text{ for some prime } p \text{ and some } m \geq 1, \\ 0 & \text{otherwise.} \end{cases}$$

   (a) Is $\Lambda(n)$ multiplicative?
   (b) Show that for any positive integer n,

$$\ln n = \sum_{d|n} \Lambda(d). \tag{2.12}$$

2. An arithmetical function $f(n)$ is *completely multiplicative* if for any positive integers m and n,

$$f(mn) = f(m)f(n).$$

   (a) Show that if $f(n)$ is a completely multiplicative arithmetical function, then the Dirichlet inverse of $f(n)$ is $\mu(n)f(n)$.
   (b) Show that if $f^{-1}(n) = \mu(n)f(n)$ for any positive integer n, then $f(n)$ is completely multiplicative.

3. (Ramanujan's sum) Let n be any positive integer. The Ramanujan sum is defined by

$$c_q(n) = \sum_{\substack{a=1 \\ (a,q)=1}}^{q} e^{2\pi i \frac{a}{q} n}.$$

   (a) Show that

$$c_q(n) = \sum_{d|(q,n)} d\mu\left(\frac{q}{d}\right).$$

   (b) Identify the function $c_q(n)$ when $n = 0$.

(c) Identify the function $c_q(n)$ when $q = 1$.

4. For real or complex α and any positive integer n, we define

$$\sigma_\alpha(n) = \sum_{d|n} d^\alpha,$$

the sum of the α-th powers of the divisors of n.

(a) Prove that $\sigma_\alpha(n)$ is multiplicative.
(b) Prove that

$$\sigma_\alpha(p^a) = \begin{cases} \dfrac{p^{\alpha(a+1)} - 1}{p^\alpha - 1} & \text{if } \alpha \neq 0 \\ a + 1 & \text{if } \alpha = 0. \end{cases}$$

(c) Find the Dirichlet inverse of $\sigma_\alpha(n)$.
(d) A number n is perfect when it is the sum of its divisors less than n. In other words, a number n is perfect if

$$\sigma(n) := \sigma_1(n) = 2n.$$

Show that if $2^p - 1$ is a prime, then $n = 2^{p-1}(2^p - 1)$ is perfect.
(e) Show that if n is an even perfect number, then n must be of the form

$$2^{k-1}(2^k - 1),$$

where $2^k - 1$ is prime.

5. (Liouville's function $\lambda(n)$) Let $\lambda(1) = 1$. Let n is any positive integer given by $n = p_1^{\alpha_1} \cdots p_k^{\alpha_k}$. Define

$$\lambda(n) = (-1)^{\alpha_1 + \cdots + \alpha_k}.$$

Show that λ is completely multiplicative and that for every $n \geq 1$,

$$\sum_{d|n} \lambda(d) = \begin{cases} 1 & \text{if } n \text{ is a square} \\ 0 & \text{otherwise.} \end{cases}$$

Using the fact that λ is completely multiplicative, find λ^{-1}.

6. Prove that for any positive integer n,

$$\sum_{d^2|n} \mu(d) = \mu^2(n).$$

7. Let $d(n)$ denotes the number of positive divisors of n. In other words,

$$d(n) = \sum_{d|n} 1 = (u * u)(n).$$

Prove that

$$\prod_{t|n} t = n^{d(n)/2}.$$

8. Find an arithmetical function $f(n)$ such that for any positive integer n,

$$\frac{1}{\varphi(n)} = \sum_{d|n} \frac{1}{d} f\left(\frac{n}{d}\right).$$

Chapter 3

Averages of Arithmetical Functions

3.1 Introduction

In this chapter, we will study the mean of an arithmetical function f. This is defined by

$$\overline{f}(n) = \frac{1}{n}\sum_{k=1}^{n} f(k).$$

The purpose of studying the arithmetic mean $\overline{f}(n)$ is because in general, $\overline{f}(n)$ behaves more regularly than $f(n)$ when n is large.

To study the average of an arbitrary function f, we need to study the partial sum

$$\sum_{k=1}^{n} f(k).$$

Sometimes it is convenient to replace the upper index n by an arbitrary positive real number x and to consider instead sums of the form

$$\sum_{k\leq x} f(k).$$

Here it is understood that the index k varies from 1 to $[x]$, where $[x]$ denotes the greatest integer less than or equal to x. If $0 < x < 1$, the sum is empty and we assign it the value 0. Our goal is to determine the behavior of this sum as a function of x, especially when x is large.

We end this introduction with the following definitions:

Definition 3.1. Let a be any real number and let $g(x)$ be a real-valued function such that $g(x) > 0$ when $x \geq a$. We write

$$f(x) = O(g(x))$$

to mean that the quotient $f(x)/g(x)$ is bounded for $x \geq a$; that is, there exists a constant $M > 0$ such that

$$|f(x)| \leq Mg(x) \quad \text{for all } x \geq a.$$

Sometimes, we will also use the notation

$$f(x) \ll g(x)$$

to represent $f(x) = O(g(x))$.

Example 3.1. The function $x^2 = O(x^3)$ when x is large. The function $x^n = O(e^x)$ for any positive integer n.

Definition 3.2. If

$$\lim_{x \to \infty} \frac{f(x)}{g(x)} = 1,$$

then we say that $f(x)$ is asymptotic to $g(x)$ as $x \to \infty$, and we write

$$f(x) \sim g(x) \quad \text{as } x \to \infty.$$

3.2 Partial summation and the Euler-Maclaurin summation formula

Theorem 3.1. *Let $a(n)$ be an arithmetic function and set*

$$A(x) = \sum_{n \leq x} a(n).$$

Let $0 \leq y < x$ be real numbers and f be a real-valued function with continuous derivative on $[y, x]$. Then

$$\sum_{y < n \leq x} a(n)f(n) = f(x)A(x) - f(y)A(y) - \int_y^x A(t)f'(t)\,dt. \tag{3.1}$$

Proof. We observe that

$$\begin{aligned}\int_y^x A(t)f'(t)\,dt &= \int_y^x \sum_{n \leq t} a(n)f'(t)\,dt \qquad (3.2)\\ &= \sum_{n \leq x} a(n) \int_{\max(y,n)}^x f'(t)\,dt\\ &= \sum_{n \leq x} a(n)[f(x) - f(\max(y,n))].\end{aligned}$$

Therefore,

$$\int_y^x A(t)f'(t)\,dt = f(x)A(x) - \sum_{n\le y} a(n)f(y) - \sum_{y<n\le x} a(n)f(n)$$
$$= f(x)A(x) - f(y)A(y) - \sum_{y<n\le x} a(n)f(n).$$

Simplifying, we find that

$$\sum_{y<n\le x} a(n)f(n) = A(x)f(x) - A(y)f(y) - \int_y^x A(t)f'(t)\,dt.$$

□

Remark 3.1. The second equality of (3.2) follows from interchanging the integral with the summation. We now explain the limits in the integral using the following diagram (see Fig. 3.1). Note that for a fixed t, the sum is over all $n \le t$ (consider the vertical line). For a fixed n, we integrate from y to x if $n < y$ and from n to x if $n \ge y$ (consider the two horizontal lines in the shaded region). Hence for a fixed n, we integrate from $\max(n, y)$ to x.

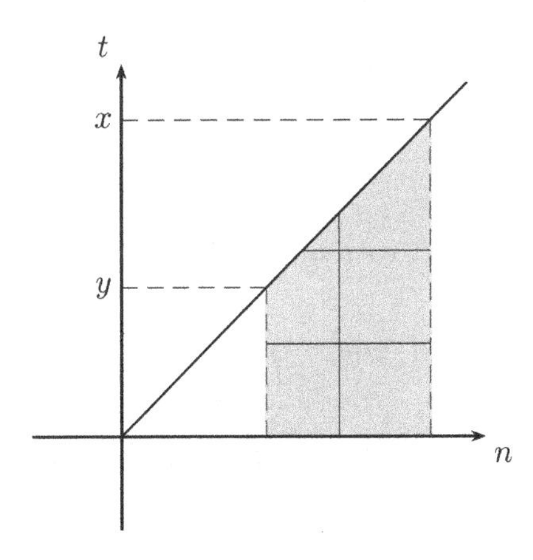

Fig. 3.1 Diagram for explaining the limits in the integral of (3.2).

When $y = 1$, we have

$$\sum_{1<n\leq x} a(n)f(n) = A(x)f(x) - A(1)f(1) - \int_1^x A(t)f'(t)\,dt.$$

But

$$\sum_{1<n\leq x} a(n)f(n) + A(1)f(1) = \sum_{1<n\leq x} a(n)f(n) + a(1)f(1)$$
$$= \sum_{n\leq x} a(n)f(n).$$

Consequently, we have

$$\sum_{n\leq x} a(n)f(n) = A(x)f(x) - \int_1^x A(t)f'(t)\,dt. \tag{3.3}$$

Theorem 3.2 (The Euler-Maclaurin summation formula). *Let x and y be positive real numbers such that $0 < y < x$ and let $f(x)$ be a real-valued function with continuous derivative on $[y, x]$. Then*

$$\sum_{y<n\leq x} f(n) = \int_y^x f(t)\,dt + \int_y^x \{t\}f'(t)\,dt - f(x)\{x\} + f(y)\{y\}.$$

Proof. By Theorem 3.1 with $a(n) = 1$ and $A(x) = [x]$, we find that

$$\sum_{y<n\leq x} f(n) = f(x)[x] - f(y)[y] - \int_y^x [t]f'(t)\,dt$$
$$= f(x)x - \{x\}f(x) + f(y)\{y\} - f(y)y - \int_y^x (t - \{t\})f'(t)\,dt$$
$$= -\{x\}f(x) + f(y)\{y\} + \int_y^x \{t\}f'(t)\,dt + f(x)x - f(y)y - \int_y^x tf'(t)\,dt$$
$$= -f(x)\{x\} + f(y)\{y\} + \int_y^x \{t\}f'(t)\,dt + \int_y^x f(t)\,dt.$$

□

3.3 Some elementary asymptotic formulas

Definition 3.3. For each real number $s > 1$, we define the Riemann zeta function as

$$\zeta(s) = \sum_{n=1}^{\infty} \frac{1}{n^s}.$$

Definition 3.4. The Euler constant C is defined as

$$C = \lim_{n\to\infty} \left(1 + \frac{1}{2} + \frac{1}{3} + \cdots + \frac{1}{n} - \ln n\right).$$

Theorem 3.3. *If $x \geq 1$, then*

(a) $\displaystyle\sum_{n\leq x} \frac{1}{n} = \ln x + C + O\left(\frac{1}{x}\right)$,

(b) $\displaystyle\sum_{n\leq x} \frac{1}{n^s} = \frac{x^{1-s}}{1-s} + C(s) + O(x^{-s})$ *if $s > 0$ and $s \neq 1$,*

where

$$C(s) = \begin{cases} \zeta(s) & \text{if } s > 0, \\ \lim\limits_{x\to\infty} \left(\displaystyle\sum_{n\leq x} \frac{1}{n^s} - \frac{x^{1-s}}{1-s}\right) & \text{if } 0 < s < 1. \end{cases}$$

(c) $\displaystyle\sum_{n>x} \frac{1}{n^s} = O(x^{1-s})$ *if $s > 1$, and*

(d) $\displaystyle\sum_{n\leq x} n^\alpha = \frac{x^{\alpha+1}}{\alpha+1} + O(x^\alpha)$ *if $\alpha \geq 0$.*

Proof. To prove (a), we first let $f(t) = 1/t$ in Theorem 3.2. Then

$$\begin{aligned}
\sum_{n\leq x} \frac{1}{n} &= \int_1^x \frac{dt}{t} - \int_1^x \frac{\{t\}}{t^2}\,dt + 1 - \frac{\{x\}}{x} \\
&= \ln x - \int_1^x \frac{\{t\}}{t^2}\,dt + 1 + O\left(\frac{1}{x}\right) \\
&= \ln x + 1 - \int_1^\infty \frac{\{t\}}{t^2}\,dt + \int_x^\infty \frac{\{t\}}{t^2}\,dt + O\left(\frac{1}{x}\right).
\end{aligned}$$

The improper integral

$$\int_1^\infty \{t\}t^{-2}\,dt$$

exists since it is dominated by

$$\int_1^\infty t^{-2}\,dt.$$

Furthermore,

$$0 \leq \int_x^\infty \frac{\{t\}}{t^2}\,dt \leq \int_x^\infty \frac{1}{t^2}\,dt = \frac{1}{x},$$

so the last equation becomes

$$\sum_{n\leq x} \frac{1}{n} = \ln x + 1 - \int_1^\infty \frac{\{t\}}{t^2}\,dt + O\left(\frac{1}{x}\right).$$

This proves (a) with

$$C = 1 - \int_1^\infty \frac{\{t\}}{t^2}\,dt = \lim_{x\to\infty}\left(\sum_{n\leq x}\frac{1}{n} - \ln x\right).$$

To prove (b), we use the same argument with

$$f(x) = x^{-s},$$

where $s > 0, s \neq 1$. The Euler-Maclaurin summation implies that

$$\sum_{n\leq x} \frac{1}{n^s} = \frac{x^{1-s}}{1-s} - \frac{1}{1-s} + 1 - s\int_1^\infty \frac{\{t\}}{t^{s+1}}\,dt + O(x^{-s}).$$

Therefore,

$$\sum_{n\leq x} \frac{1}{n^s} = \frac{x^{1-s}}{1-s} + C(s) + O(x^{-s}), \tag{3.4}$$

where

$$C(s) = 1 - \frac{1}{1-s} - s\int_1^\infty \frac{\{t\}}{t^{s+1}}\,dt.$$

If $s > 1$ then the left-hand side of (3.4) approaches $\zeta(s)$ as x approaches ∞ and both x^{1-s} and x^{-s} approach 0. Hence

$$C(s) = \zeta(s)$$

if $s > 1$. If $0 < s < 1$, then

$$\lim_{x\to\infty} \frac{1}{x^s} = 0$$

and (3.4) shows that

$$C(s) = \lim_{x\to\infty} \left(\sum_{n\le x} \frac{1}{n^s} - \frac{x^{1-s}}{1-s} \right)$$

and this completes the proof of (b).

To prove (c), we use (b) with $s > 1$ to obtain

$$\sum_{n>x} \frac{1}{n^s} = \zeta(s) - \sum_{n\le x} \frac{1}{n^s} = \frac{x^{1-s}}{s-1} + O(x^{-s}) = O(x^{1-s})$$

since $x^{-s} \le x^{1-s}$.

Finally, to prove (d), we use the Euler-Maclaurin summation formula with $f(t) = t^\alpha$ to obtain

$$\begin{aligned}\sum_{n\le x} n^\alpha &= \int_1^x t^\alpha\,dt + \alpha \int_1^x t^{\alpha-1}\{t\}\,dt + 1 - \{x\}x^\alpha \\ &= \frac{x^{\alpha+1}}{\alpha+1} - \frac{1}{\alpha+1} + O\left(\alpha \int_1^x t^{\alpha-1}\,dt\right) + O(x^\alpha) \\ &= \frac{x^{\alpha+1}}{\alpha+1} + O(x^\alpha).\end{aligned}$$

□

3.4 The divisor function and Dirichlet's hyperbola method

Definition 3.5. Let $d(1) = 1$ and for positive integer $n > 1$, we define $d(n)$ to be the number of divisors of n.

We note that

$$d(n) = \sum_{d|n} 1$$

and so,

$$d = u * u, \tag{3.5}$$

where u is defined in Definition 2.10. In this section, we derive Dirichlet's asymptotic formula for the partial sums of the divisor function $d(n)$.

Theorem 3.4. *Let f and g be two arithmetic functions with*

$$F(x) = \sum_{n\le x} f(n), \quad \text{and} \quad G(x) = \sum_{n\le x} g(n).$$

For $1 \le y \le x$, we have

$$\sum_{n\le x} (f*g)(n) = \sum_{n\le y} g(n) F\left(\frac{x}{n}\right) + \sum_{m\le \frac{x}{y}} f(m) G\left(\frac{x}{m}\right) - F\left(\frac{x}{y}\right) G(y).$$

Proof. First, we observe that

$$\sum_{n\le x}(f*g)(n)=\sum_{md\le x}f(m)g(d).$$

Next, for $y\le x$, we find that

$$\begin{aligned}\sum_{md\le x}f(m)g(d)&=\sum_{\substack{md\le x\\ d\le y}}f(m)g(d)+\sum_{\substack{md\le x\\ d>y}}f(m)g(d)\\ &=\sum_{d\le y}g(d)F\left(\frac{x}{d}\right)+\sum_{m\le\frac{x}{y}}f(m)\left\{G\left(\frac{x}{m}\right)-G(y)\right\}.\end{aligned}$$

□

We now set $f=g=u$. Then

$$f*g=u*u=d$$

by (3.5). Note that $F(x)=[x]=G(x)$. Let $y=\sqrt{x}$. Then by Theorem 3.4,

$$\begin{aligned}\sum_{n\le x}d(n)&=2\sum_{n\le\sqrt{x}}\left[\frac{x}{n}\right]-[\sqrt{x}]^2\\ &=2x\sum_{n\le\sqrt{x}}\frac{1}{n}-x+O(\sqrt{x}).\end{aligned}$$

Using Theorem 3.3 (a), we conclude that

Theorem 3.5. *For all* $x\ge 1$,

$$\sum_{n\le x}d(n)=x\ln x+(2\gamma-1)x+O(\sqrt{x}),$$

where γ *is the Euler's constant.*

As a corollary, we deduce that

$$\overline{d}(n)\sim\ln n. \tag{3.6}$$

In other words, the average order of $d(n)$ is $\ln n$.

The asymptotic (3.6) can be shown without using the Dirichlet's Hyperbola method. However, the error term obtained would be $O(x)$ instead of $O(\sqrt{x})$ (see Problem 2 of Exercise 3.6).

Remark 3.2. The error term in Theorem 3.5 can be improved. In 1903 Voronoi [13] proved that it is $O(x^{1/3}\log x)$. In 1928, J.G. van der Corput [12] improved the error term to $O(x^{27/82})$ using the method of exponential sums. In 1988, H. Iwaniec and C.J. Mozzochi [7] showed that the error term can be taken as $O(x^{7/22})$. The best possible error term is one given recently by M.N. Huxley in 2003 [6], who showed that the error is $O(x^{131/416}(\ln x)^{26947/8320})$.

3.5 An application of the hyperbola method

An interesting question one can ask is:

"If two positive integers are randomly chosen, what is the probability that they are relatively prime?"

To answer this question, we first show the following result:

Theorem 3.6. *Let $\varphi(n)$ be the Euler φ function. For $x > 1$,*

$$\sum_{n\le x}\varphi(n) = x^2\frac{3}{\pi^2} + O(x^{3/2}).$$

Proof. We recall that $\varphi = \mu * N$. Applying Theorem 3.4 with $f = N$ and $g = \mu$, we find that

$$\sum_{n\le x}\varphi(n) = \sum_{n\le x}\mu * N(n) = \sum_{n\le y}\mu(n)F\left(\frac{x}{n}\right) + \sum_{m\le \frac{x}{y}} N(m)G\left(\frac{x}{m}\right) - F\left(\frac{x}{y}\right)G(y),$$

where

$$F(x) = \sum_{n\le x} N(n) = \frac{x^2}{2} + O(x)$$

and

$$G(x) = \sum_{n\le x}\mu(n) = O(x).$$

Therefore,

$$\sum_{n\le x}\varphi(n) = \frac{1}{2}\sum_{n\le y}\mu(n)\left(\frac{x}{n}\right)^2 + O\left(\left|\sum_{n\le y}\mu(n)\right| x\right)$$
$$+ O\left(\sum_{m\le \frac{x}{y}} m\frac{x}{m}\right) + O\left(\left(\frac{x}{y}\right)^2 y\right).$$

Let $y = \sqrt{x}$ and we conclude that

$$\sum_{n\le x}\varphi(n) = \sum_{n\le\sqrt{x}}\frac{\mu(n)}{n^2}\frac{x^2}{2} + O(x^{3/2}). \tag{3.7}$$

It will be shown in Theorem 6.3 that

$$\zeta(2)\sum_{n=1}^{\infty}\frac{\mu(n)}{n^2} = 1.$$

This implies that

$$\sum_{n=1}^{\infty} \frac{\mu(n)}{n^2} = \frac{6}{\pi^2},$$

since (see [1, p. 190, Ex. 1])

$$\zeta(2) = \frac{\pi^2}{6}.$$

Given the above identity, we find that

$$\begin{aligned}\sum_{n\le\sqrt{x}} \frac{\mu(n)}{n^2} &= \sum_{n=1}^{\infty} \frac{\mu(n)}{n^2} - \sum_{n>\sqrt{x}} \frac{\mu(n)}{n^2} \\ &= \sum_{n=1}^{\infty} \frac{\mu(n)}{n^2} + O\left(\sum_{n>\sqrt{x}} \frac{1}{n^2}\right) \\ &= \frac{6}{\pi^2} + O(x^{-1/2}),\end{aligned}$$

where the last equality follows from Theorem 3.3 (c). Substituting the above into (3.7), we conclude the proof of the theorem. □

Now, let T be a positive integer and

$$S = \{(m,n)|1 \le m \le T, 1 \le n \le T\}.$$

Then the total number of elements in S such that $(m,n) = 1$ is given by

$$\begin{aligned}\sum_{\substack{n\le T}} \sum_{\substack{m\le T \\ (m,n)=1}} 1 &= 1 + 2\sum_{m\le T} \sum_{\substack{n<m \\ (m,n)=1}} 1 \\ &= 1 + 2\sum_{m\le T} \varphi(m) = \frac{6}{\pi^2}T^2 + O(T^{3/2}).\end{aligned}$$

This shows that the probability that two randomly chosen positive integers are relatively prime is

$$\lim_{T\to\infty} \frac{|S|}{T^2} = \frac{6}{\pi^2}.$$

3.6 Exercises

1. Use the Euler-Maclaurin summation formula to show that for $x \geq 3$,

$$\sum_{n\leq x} (\ln n)^2 = x\ln^2 x - 2x\ln x + 2x + O(\ln^2 x).$$

2. Show that for any arithmetical function $f(n)$,

$$\sum_{n\leq x}\sum_{d|n} f(d) = \sum_{d\leq x} f(d)\left[\frac{x}{d}\right]. \tag{3.8}$$

   Deduce that

$$\sum_{n\leq x} d(n) = x\ln x + O(x).$$

3. Use partial summation formula to determine $f(x)$ if

$$\sum_{n\leq x} \frac{d(n)}{n} \sim f(x).$$

4. **Definition 3.6.** Let n be any positive integer. Let $\omega(n)$ be the number of distinct prime factors of n. More precisely,

$$\omega(n) = \begin{cases} 0 & \text{if } n = 1, \\ k & \text{if } n = p_1^{\alpha_1} p_2^{\alpha_2} \cdots p_k^{\alpha_k}. \end{cases}$$

   Show that for $x \geq 3$,

$$\sum_{n\leq x} 2^{\omega(n)} = \frac{6}{\pi^2} x\ln x + O(x).$$

5. Let $x \geq 1$. Prove that

$$\sum_{n\leq x} \varphi(n) = \frac{1}{2}\sum_{n\leq x} \mu(n)\left[\frac{x}{n}\right]^2 + \frac{1}{2}.$$

   Hence, deduce that for $x \geq 2$,

$$\sum_{n\leq x} \varphi(n) = \frac{1}{2}\frac{x^2}{\zeta(2)} + O(x\ln x).$$

6. Let $x \geq 1$. Prove that

$$\sum_{n \leq x} \frac{\varphi(n)}{n} = \sum_{n \leq x} \frac{\mu(n)}{n} \left[\frac{x}{n}\right].$$

Hence, deduce that for $x \geq 2$,

$$\sum_{n \leq x} \frac{\varphi(n)}{n} = \frac{x}{\zeta(2)} + O(\ln x).$$

7. Let n be any positive integer and

$$\sigma(n) = \sum_{d|n} d.$$

Assume that

$$\zeta(2) = \pi^2/6.$$

(a) Show that if $n \geq 2$, then

$$\frac{\sigma(n)}{n} < \frac{n}{\varphi(n)} < \frac{\pi^2}{6} \frac{\sigma(n)}{n}.$$

(b) Let $x \geq 2$. Prove that

$$\sum_{n \leq x} \frac{n}{\varphi(n)} = O(x).$$

Chapter 4

Elementary Results on the Distribution of Primes

4.1 Introduction

Definition 4.1. For real number $x > 0$, let $\pi(x)$ denote the number of primes not exceeding x.

The behavior of $\pi(x)$ as the function of x has been studied by many mathematicians ever since the eighteenth century. Inspection of tables of primes led C.F. Gauss (1792) and A.M. Legendre (1798) to conjecture that

$$\pi(x) \sim \frac{x}{\ln x}. \tag{4.1}$$

This conjecture was first proved independently by J. Hadamard and de la Vallée Poussin in 1896 and is now known as the *Prime Number Theorem.* We record the theorem as follows:

Theorem 4.1 (Prime Number Theorem). *Let x be a real positive number and $\pi(x)$ be the number of primes less than x. Then*

$$\pi(x) \sim \frac{x}{\ln x}.$$

Proofs of the Prime Number Theorem are often classified as elementary or analytic. The proofs of J. Hadamard and de la Vallee Poussin are analytic, using complex function theory and properties of the Riemann zeta function $\zeta(s)$ (see Definition 3.3 for the definition of $\zeta(s)$ when $s \in \mathbf{R}$ and $s > 1$). Elementary proofs were discovered around 1949 by A. Selberg and P. Erdös. Their proofs do not involve $\zeta(s)$ and complex function theory, hence the name "elementary".

There are other elementary proofs of the prime number theorem since the appearance of the work of Selberg and Erdös, one of which is [4]. The proof given in [4] relies on proving an equivalent statement of the Prime Number Theorem and the mean value of $\mu(n)$. This equivalent statement of the Prime Number Theorem will be established in Section 4.5.

In this chapter, we derive some basic properties of $\pi(x)$ and establish several statements equivalent to the Prime Number Theorem. We will also use the results discussed in this chapter to study Bertrand's Postulate, which states that for $n \geq 2$, there exists a prime between n and $2n$.

4.2 The function $\psi(x)$

We recall the definition of Mangoldt's function (see Exercise 2.6, Problem 1).

Definition 4.2. Let n be a positive integer and let

$$\Lambda(n) = \begin{cases} \ln p, & \text{if } n \text{ is a prime power} \\ 0, & \text{otherwise.} \end{cases}$$

Definition 4.3. For real number $x \geq 1$,

$$\psi(x) = \sum_{n \leq x} \Lambda(n) = \sum_{p^m \leq x} \ln p.$$

Theorem 4.2. *There exist positive constants c_1 and c_2 such that*

$$c_1 x \leq \psi(x) \leq c_2 x.$$

Proof. For $x \geq 4$, let

$$S = \sum_{n \leq x} \ln n - 2 \sum_{n \leq \frac{x}{2}} \ln n.$$

By Theorem 3.2 with $f(n) = \ln n$, we find that

$$\sum_{n \leq x} \ln n = \int_1^x \ln t dt + \int_1^x \{t\} \frac{1}{t} dt - \{x\} \ln x + \{y\} \ln y$$
$$= x \ln x - x + O(\ln x). \tag{4.2}$$

This implies that

$$S = x \ln 2 + O(\ln x).$$

Therefore, there exists an $x_0 \geq 4$ such that

$$\frac{x}{2} \leq S \leq x \tag{4.3}$$

whenever $x \geq x_0 \geq 4$. Next, since (see Section 2.6, Problem 1)

$$\ln n = \sum_{d|n} \Lambda(d),$$

we find that

$$\begin{aligned} S &= \sum_{n \leq x} \sum_{d|n} \Lambda(d) - 2 \sum_{n \leq \frac{x}{2}} \sum_{d|n} \Lambda(d) \\ &= \sum_{d \leq x} \Lambda(d) \left[\frac{x}{d}\right] - 2 \sum_{d \leq \frac{x}{2}} \Lambda(d) \left[\frac{x}{2d}\right] \\ &= \sum_{d \leq \frac{x}{2}} \Lambda(d) \left\{\left[\frac{x}{d}\right] - 2\left[\frac{x}{2d}\right]\right\} + \sum_{\frac{x}{2} < d \leq x} \Lambda(d) \left[\frac{x}{d}\right]. \end{aligned}$$

Hence,

$$S = \sum_{d \leq \frac{x}{2}} \Lambda(d)\theta_d + \sum_{\frac{x}{2} < d \leq x} \Lambda(d) \left[\frac{x}{d}\right],$$

where

$$\theta_d = \left[\frac{x}{d}\right] - 2\left[\frac{x}{2d}\right]. \tag{4.4}$$

Now, for

$$\frac{x}{2} < d \leq x,$$

we have

$$\left[\frac{x}{d}\right] = 1.$$

Therefore, we may simplify the second term on the right-hand side of (4.4) to obtain

$$S = \sum_{d \leq \frac{x}{2}} \Lambda(d)\theta_d + \sum_{\frac{x}{2} < d \leq x} \Lambda(d). \tag{4.5}$$

We now observe that $\theta_d = 0$ or 1 since

$$[y] - 2[y/2] = 0 \quad \text{or} \quad 1.$$

Using (4.5), we deduce that

$$S \leq \sum_{d \leq \frac{x}{2}} \Lambda(d) + \sum_{\frac{x}{2} < d \leq x} \Lambda(d) = \sum_{d \leq x} \Lambda(d) = \psi(x) \tag{4.6}$$

and

$$S \geq \sum_{\frac{x}{2} < d \leq x} \Lambda(d) = \psi(x) - \psi\left(\frac{x}{2}\right). \tag{4.7}$$

From (4.3) and (4.6),

$$\psi(x) \geq S \geq \frac{x}{2} \quad (x \geq x_0).$$

Therefore,

$$\psi(x) \geq c_1 x.$$

To obtain a lower bound for $\psi(x)$, we first deduce from (4.3), (4.7) that

$$\psi(x) - \psi\left(\frac{x}{2}\right) \leq S \leq x.$$

Therefore,

$$\begin{aligned}
\psi(x) &\leq x + \psi\left(\frac{x}{2}\right), \quad x \geq x_0 \\
&\leq x + \frac{x}{2} + \psi\left(\frac{x}{4}\right), \quad x \geq 2x_0 \\
&\vdots \\
&\leq x + \frac{x}{2} + \cdots + \frac{x}{2^k} + \psi\left(\frac{x}{2^{k+1}}\right), \quad \frac{x}{2^{k+1}} < x_0 \leq \frac{x}{2^k}.
\end{aligned}$$

This implies that

$$\psi(x) \leq 2x + \psi(x_0) \leq c_2 x$$

for some positive real number c_2. □

4.3 The functions $\theta(x)$ and $\pi(x)$

Definition 4.4. For real number $x \geq 1$, let

$$\theta(x) = \sum_{p \leq x} \ln p.$$

Theorem 4.3. *For real number $x \geq 1$, we have*

$$\theta(x) = \psi(x) + O(\sqrt{x}).$$

Proof. We first note that the difference of $\psi(x)$ and $\theta(x)$ is

$$\begin{aligned}
\psi(x) - \theta(x) &= \sum_{\substack{p^m \leq x \\ m \leq 2}} \ln p \\
&= \sum_{\substack{p \leq \sqrt{x} \\ m = 2}} \ln p + \sum_{p \leq x^{1/3}} \ln p \sum_{3 \leq m \leq \frac{\ln x}{\ln p}} 1.
\end{aligned}$$

Hence,

$$\begin{aligned}\psi(x) - \theta(x) &\leq \psi(\sqrt{x}) + \sum_{p \leq x^{1/3}} \ln p \frac{\ln x}{\ln p} \\ &\ll \sqrt{x} + x^{1/3} \ln x \ll \sqrt{x},\end{aligned}$$

where $f(x) \ll g(x)$ is another notation for $f(x) = O(g(x))$ (see Definition 3.1). □

Using Theorems 4.2 and 4.3, we deduce the following corollary.

Corollary 4.4. *For $x \geq 4$, there exist real positive constants c_1 and c_2 such that*

$$c_1 x \leq \theta(x) \leq c_2 x.$$

We give a relation between $\theta(x)$ and $\pi(x)$, where $\pi(x)$ is given by Definition 4.1.

Theorem 4.5. *For each positive real $x \geq 4$,*

$$\frac{c_1 x}{\ln x} \leq \pi(x) \leq \frac{c_2 x}{\ln x}.$$

Proof. It suffices to prove that

$$\pi(x) = \frac{1}{\ln x}\theta(x) + O\left(\frac{x}{\ln^2 x}\right)$$

by Theorem 4.3. We observe that

$$\begin{aligned}\pi(x) - \frac{\theta(x)}{\ln x} &= \sum_{p \leq x} \left(1 - \frac{\ln p}{\ln x}\right) \\ &= \sum_{p \leq x} \ln p \left(\frac{1}{\ln p} - \frac{1}{\ln x}\right).\end{aligned} \tag{4.8}$$

If

$$a(n) = \begin{cases} \ln p & \text{if } n \text{ is a prime } p, \\ 0 & \text{otherwise,} \end{cases}$$

then by Corollary 4.4,

$$A(t) = \sum_{n \leq t} a(n) = \theta(t) \ll t.$$

The last expression in (4.8) is

$$\begin{aligned}\theta(x)\left(\frac{1}{\ln x}-\frac{1}{\ln x}\right)-\int_2^x \theta(t)\left(\frac{1}{\ln t}-\frac{1}{\ln x}\right)' dt &= \int_2^x \frac{\theta(t)}{t\ln^2 t}\,dt \ll \int_2^x \frac{dt}{\ln^2 t}\\ &= \int_2^{\sqrt{x}} \frac{dt}{\ln^2 t}+\int_{\sqrt{x}}^x \frac{dt}{\ln^2 t}\\ &\ll \sqrt{x}+\int_{\sqrt{x}}^x \frac{dt}{\ln^2 x} \ll \frac{x}{\ln^2 x}.\end{aligned}$$

□

As corollaries of Theorems 4.3 and 4.5, we have the following results. We leave the details of the proofs of these corollaries to the readers.

Corollary 4.6. *The Prime Number Theorem*

$$\pi(x) \sim \frac{x}{\ln x}$$

is equivalent to each of the following relations:

(a) $\theta(x) \sim x$, *and*
(b) $\psi(x) \sim x$.

4.4 Merten's estimates

In this section, we show that there are infinitely many primes by showing that $\sum_{p\le x}\frac{1}{p}$ diverges.

Theorem 4.7 (Merten's estimates). *Let x be a positive real number greater than 1. We have*

(a) $\displaystyle\sum_{n\le x}\frac{\Lambda(n)}{n} = \ln x + O(1)$,
(b) $\displaystyle\sum_{p\le x}\frac{\ln p}{p} = \ln x + O(1)$,
(c) $\displaystyle\sum_{p\le x}\frac{1}{p} = \ln\ln x + A + O\left(\frac{1}{\ln x}\right)$, *and*
(d) *(Merten's Theorem)* $\displaystyle\prod_{p\le x}\left(1-\frac{1}{p}\right) = \frac{e^{-A}}{\ln x}\left(1+O\left(\frac{1}{\ln x}\right)\right)$,

where A is a constant.

Proof.
(a) First, we write

$$\sum_{n \le x} \frac{\Lambda(n)}{n} = \sum_{n \le x} \left\{ \Lambda(n) \frac{1}{x} \left(\left[\frac{x}{n} \right] + O(1) \right) \right\}$$
$$= \frac{1}{x} \sum_{n \le x} \Lambda(n) \left[\frac{x}{n} \right] + O\left(\frac{1}{x} \sum_{n \le x} \Lambda(n) \right).$$

By (3.8), we find that

$$\sum_{n \le x} \Lambda(n) \left[\frac{x}{n} \right] = \sum_{n \le x} (\Lambda * u)(n).$$

Hence, we deduce using (2.12) and (4.2) that

$$\sum_{n \le x} \frac{\Lambda(n)}{n} = \frac{1}{x} \sum_{n \le x} (\Lambda * u)(n) + O(1)$$
$$= \frac{1}{x} \sum_{n \le x} \ln n + O(1),$$
$$= \ln x + O(1).$$

(b) We observe that

$$0 \le \sum_{n \le x} \frac{\Lambda(n)}{n} - \sum_{p \le x} \frac{\ln p}{p} = \sum_{p \le \sqrt{x}} \ln p \sum_{2 \le m \le \frac{\ln x}{\ln p}} \frac{1}{p^m}$$
$$\ll \sum_{p \le \sqrt{x}} \frac{\ln p}{p^2} \ll 1.$$

Hence,

$$\sum_{p \le x} \frac{\ln p}{p} = \ln x + O(1).$$

(c) Let

$$A(x) = \sum_{n \le x} a(n)$$

where

$$a(n) = \begin{cases} \dfrac{\ln p}{p}, & \text{if } p \text{ is prime} \\ 0, & \text{otherwise.} \end{cases}$$

Then, we find that

$$\begin{aligned}\sum_{p\le x}\frac{1}{p}&=\sum_{p\le x}\left(\frac{\ln p}{p}\right)\left(\frac{1}{\ln p}\right)\\&=A(x)\frac{1}{\ln x}-\int_2^x A(t)\left(\frac{1}{\ln t}\right)'\,dt\\&=\frac{A(x)}{\ln x}+\int_2^x\frac{A(t)}{t\ln^2 t}\,dt.\end{aligned}\tag{4.9}$$

By Theorem 4.7 (b), we find that

$$A(t)=\ln t+R(t),$$

with

$$R(t)\ll 1,\quad t\ge 2.\tag{4.10}$$

Using (4.10) in the last term of (4.9), we deduce that

$$\begin{aligned}\int_2^x\frac{\ln t+R(t)}{t\ln^2 t}dt&=\int_2^x\frac{dt}{t\ln t}+\int_2^x\frac{R(t)}{t\ln^2 t}dt\\&=\ln\ln x-\ln\ln 2+\int_2^\infty\frac{R(t)}{t\ln^2 t}dt-\int_x^\infty\frac{R(t)}{t\ln^2 t}\,dt\\&=\ln\ln x-\ln\ln 2+A''+O\left(\frac{1}{\ln x}\right).\end{aligned}\tag{4.11}$$

Substituting (4.11) into (4.9), we conclude our proof of (c).

(d) We observe that

$$\begin{aligned}\ln\prod_{p\le x}\left(1-\frac{1}{p}\right)&=\sum_{p\le x}\ln\left(1-\frac{1}{p}\right)\\&=\sum_{p\le x}\left(-\frac{1}{p}+r_p\right),\end{aligned}$$

where

$$r_p=\ln\left(1-\frac{1}{p}\right)+\frac{1}{p}.$$

Hence,

$$\begin{aligned}\ln\prod_{p\le x}\left(1-\frac{1}{p}\right)&=\sum_{p\le x}r_p-\sum_{p\le x}\frac{1}{p}\\&=-\ln\ln x+A+O\left(\frac{1}{\ln x}\right)+\sum_{p}r_p-\sum_{p>x}r_p.\end{aligned}\tag{4.12}$$

Now,

$$r_p = -\sum_{m=2}^{\infty} \frac{1}{mp^m} = O\left(\frac{1}{p^2}\right), \tag{4.13}$$

since for $m \geq 1$ and $p \geq 2$,

$$mp^m \geq 2^m.$$

Using (4.13) in (4.12), we deduce that

$$\begin{aligned}
\ln \prod_{p \leq x} \left(1 - \frac{1}{p}\right) &= -\ln \ln x + A' + O\left(\frac{1}{\ln x}\right) + O\left(\sum_{p>x} \frac{1}{p^2}\right) \\
&= -\ln \ln x + A' + O\left(\frac{1}{\ln x}\right) + O\left(\frac{1}{x-1}\right) \\
&= -\ln \ln x + A' + O\left(\frac{1}{\ln x}\right).
\end{aligned}$$

Hence,

$$\ln \prod_{p \leq x} \left(1 - \frac{1}{p}\right) = -\ln \ln x + A' + O\left(\frac{1}{\ln x}\right). \tag{4.14}$$

Exponentiating both sides of (4.14), we arrive at

$$\begin{aligned}
\prod_{p \leq x} \left(1 - \frac{1}{p}\right) &= \exp\left(-\ln \ln x + A' + O\left(\frac{1}{\ln x}\right)\right) \\
&= \frac{e^{A'}}{\ln x} \exp\left(O\left(\frac{1}{\ln x}\right)\right) \\
&= \frac{e^{A'}}{\ln x} \left(1 + O\left(\frac{1}{\ln x}\right)\right),
\end{aligned}$$

since $e^t = 1 + O(t)$. □

4.5 Prime Number Theorem and $M(\mu)$

Definition 4.5. Let f be an arithmetical function. We define

$$M(f) = \lim_{x \to \infty} \frac{1}{x} \sum_{n \leq x} f(n)$$

if the limit on the right-hand side exists.

As mentioned in the introduction, there are several elementary proofs of the Prime Number Theorem. One of the proofs relies on showing that $M(\mu) = 0$ (see [4]). In this section, we will show that if $M(\mu) = 0$ then the Prime Number Theorem is true. Conversely, the Prime Number Theorem implies that $M(\mu) = 0$.

Theorem 4.8. *The Prime Number Theorem is equivalent to the relation*

$$M(\mu) = 0.$$

Proof. We will first show that the Prime Number Theorem implies that $M(\mu) = 0$.

Define

$$M_1(\mu) = \lim_{x\to\infty}\left(\frac{1}{x\ln x}\sum_{n\le x}\mu(n)\ln n\right).$$

Note that

$$M(\mu) = 0 \text{ if and only if } M_1(\mu) = 0, \tag{4.15}$$

since

$$\frac{1}{x}\sum_{n\le x}\left|\frac{\ln n}{\ln x} - 1\right| \ll \frac{1}{\ln x}.$$

Assume the Prime Number Theorem in the form $\theta(x) \sim x$.

We observe that

$$\begin{aligned}\sum_{n\le x}\mu(n)\ln n &= \sum_{n\le x}\mu(n)\sum_{d|n}\Lambda(d)\\ &= \sum_{n\le x}\mu(n)\sum_{p|n}\ln p,\end{aligned}$$

where we have used the fact that the terms with n non-squarefree are 0. Furthermore, the value $\Lambda(d)$ is nonzero only when d is a prime power. But since n is squarefree, the divisors $d|n$ that are prime powers are simply primes. Hence,

$$\begin{aligned}\sum_{n\le x}\mu(n)\ln n &= \sum_{p\le x}\ln p\sum_{\substack{n\le x\\ p|n}}\mu(n)\\ &= \sum_{p\le x}\ln p\sum_{\substack{n'\le x/p\\ p\nmid n'}}(\mu(n')),\end{aligned}$$

where we have written $n = pn'$. Next, observe that

$$\begin{aligned}
\sum_{n\le x}\mu(n)\ln n &= -\sum_{p\le x}\ln p\sum_{n\le x/p}\mu(n) + O\left(\sum_{p\le x}\ln p\sum_{\substack{n\le x/p\\ p|n}}1\right)\\
&= -\sum_{p\le x}\ln p\sum_{n\le x/p}\mu(n) + O\left(x\sum_{p\le x}\frac{\ln p}{p^2}\right)\\
&= -\sum_{n\le x}\mu(n)\sum_{p\le x/n}\ln p + O(x)\\
&= -\sum_{n\le x}\mu(n)\theta\left(\frac{x}{n}\right) + O(x).
\end{aligned}$$

Now, write

$$\sum_{n\le x}\mu(n)\ln n = -x\sum_{n\le x}\frac{\mu(n)}{n} - \sum_{n\le x}\mu(n)R\left(\frac{x}{n}\right) + O(x), \tag{4.16}$$

where

$$R(y) = \theta(y) - y.$$

Since

$$1 = \sum_{n\le x}\sum_{d|n}\mu(d) = \sum_{n\le x}\mu(n)\left[\frac{x}{n}\right] = x\sum_{n\le x}\frac{\mu(n)}{n} + O(x),$$

we find that

$$x\sum_{n\le x}\frac{\mu(n)}{n} = O(x).$$

Therefore, (4.16) may be written as

$$\sum_{n\le x}\mu(n)\ln n = -\sum_{n\le x}\mu(n)R\left(\frac{x}{n}\right) + O(x).$$

This implies that

$$\frac{1}{x\ln x}\left|\sum_{n\le x}\mu(n)\ln n\right| \le \frac{1}{x\ln x}\sum_{n\le x}\left|R\left(\frac{x}{n}\right)\right| + O\left(\frac{1}{\ln x}\right).$$

Let $\epsilon > 0$ be given. By the Prime Number Theorem,

$$\lim_{y\to\infty}\frac{R(y)}{y} = 0.$$

Therefore, there exists a y_0 such that

$$|R(y)| \leq \epsilon y \quad (y \geq y_0).$$

For $x \geq y_0$,

$$\begin{aligned}\sum_{n\leq x}\left|R\left(\frac{x}{n}\right)\right| &\leq \sum_{n\leq x/y_0} \epsilon\frac{x}{n} + \sum_{x/y_0<n\leq x} \max_{y\leq y_0}|R(y)| \\ &\leq \epsilon x \ln x + O_\epsilon(x).\end{aligned}$$

Thus,

$$\overline{\lim}\, \frac{1}{x\ln x}\left(\sum_{n\leq x}\left|R\left(\frac{x}{n}\right)\right|\right) \leq \epsilon.$$

This shows that the Prime Number Theorem implies that $M_1(\mu) = 0$. By (4.15), we deduce that $M(\mu) = 0$.

To prove the converse, let

$$\ln n = d(n) - 2C + r(n)$$

where $r(n)$ is some arithmetical function. Let $y \geq 1$. By (4.2),

$$\begin{aligned}\sum_{n\leq y} r(n) &= \sum_{n\leq y}\ln n - \sum_{n\leq y} d(n) + 2Cy + O(1) \\ &= y(\ln y - 1) + O(\ln y) \\ &\quad - (y\ln y + (2C-1)y + O(\sqrt{y})) + 2Cy + O(1) = O(\sqrt{y}).\end{aligned} \tag{4.17}$$

Next, since

$$\Lambda = \ln *\mu,$$

we conclude that

$$\begin{aligned}\sum_{n\leq x}\Lambda(n) &= \sum_{n\leq x}(\mu * \ln)(n) \\ &= \sum_{n\leq x}(\mu * d) - 2C(\sum_{n\leq x}\mu * u) + \sum_{n\leq x}(\mu * r) \\ &= [x] - 2C + \sum_{n\leq x}(\mu * r)(n).\end{aligned}$$

The last equality holds because

$$\sum_{n\leq x}\mu * u * u = [x].$$

Thus, the Prime Number Theorem follows from $\psi(x) \sim x$ (see Corollary 4.6) if

$$\lim_{x\to\infty} \frac{1}{x} \sum_{n\le x} (\mu * r)(n) = 0.$$

Now, by Theorem 3.4,

$$\begin{aligned}\sum_{n\le x} \mu * r(n) &= \sum_{n\le x} \sum_{d_1 d_2 = n} \mu(d_1) r(d_2) \\ &= \sum_{d_1\le x} \sum_{\substack{d_2\le x\\ d_1 d_2\le x}} \mu(d_1) r(d_2) \\ &= S_1 + S_2 - S_3,\end{aligned}$$

where

$$\begin{aligned} S_1 &= \sum_{d_2\le y} \sum_{d_1\le x/d_2} \mu(d_1) r(d_2) \\ S_2 &= \sum_{d_1\le x/y} \sum_{d_2\le x/d_1} \mu(d_1) r(d_2) \\ S_3 &= \sum_{d_1\le x/y} \sum_{d_2\le y} \mu(d_1) r(d_2), \end{aligned}$$

and y is a parameter in $[1, x]$ to be chosen.

For a fixed $y \in [1, x]$,

$$|S_1| \le \sum_{d_2\le y} |r(d_2)| \left| \sum_{d_1\le x/d_2} \mu(d_1) \right|.$$

Now, using the assumption that

$$M(\mu) = 0,$$

we find that

$$\lim_{x\to\infty} \left| \frac{1}{x} \sum_{d_1\le x/d_2} \mu(d_1) \right| = 0.$$

Next, using (4.17), we deduce that

$$\begin{aligned}|S_2| &\le \sum_{d_1\le x/y} \left| \sum_{d_2\le x/d_1} r(d_2) \right| \le c \sum_{d_1\le x/y} \sqrt{\frac{x}{d_1}} \\ &\le c\sqrt{x} \sum_{d_1\le x/y} \frac{1}{\sqrt{d_1}} \le c\sqrt{x} \left(1 + \int_1^{x/y} \frac{dt}{\sqrt{t}}\right) \le c_1 \frac{x}{\sqrt{y}}.\end{aligned}$$

Finally, by using (4.17), we deduce that

$$|S_3| \leq \frac{x}{y}\left|\sum_{d_2 \leq y} r(d_2)\right| \leq \frac{x}{y}c_2\sqrt{y} = c_2\frac{x}{\sqrt{y}}.$$

Hence,

$$\overline{\lim_{x\to\infty}}\ \frac{1}{x}\left|\sum_{n\leq x}(\mu * r)(n)\right| \leq 0 + c_1\frac{1}{\sqrt{y}} + \frac{c_2}{\sqrt{y}}.$$

Since y is arbitrary,

$$\lim_{x\to\infty}\frac{1}{x}\sum_{n\leq x}(\mu * r)(n) = 0.$$

□

4.6 The Bertrand Postulate

In this section, we will use the properties of the functions $\theta(x)$ and $\psi(x)$ to give a proof of the well-known Bertrand's Postulate.

Theorem 4.9 (Bertrand's Postulate). *Let n be an integer. Then for $n \geq 2$, there exists a prime p between n and $2n$.*

Most books that discuss Theorem 4.9 prove the result following Erdös' approach (see Exercise 4.7, Problem 8). In this book, we present the proof due to S. Ramanujan. [9] This proof was mentioned in an interesting article by P. Erdös titled "Ramanujan and I" [3]. Erdös' proof of Theorem 4.9 was published around 1932 and it was Kalmar who asked Erdös to look up on Ramanujan's proof and that was the first time Erdös heard about Ramanujan [3].

By definitions of $\psi(x)$ and $\theta(x)$, we observe that

Lemma 4.10. *For each positive real number x,*

$$\psi(x) = \theta(x) + \theta(\sqrt{x}) + \theta(\sqrt[3]{x}) + \cdots. \tag{4.18}$$

Next, we will show that

Lemma 4.11.

$$\ln([x]!) = \psi(x) + \psi\left(\frac{x}{2}\right) + \psi\left(\frac{x}{3}\right) + \cdots. \tag{4.19}$$

Proof. The function

$$\psi(x) = \sum_{n \le x} \Lambda(n),$$

where $\Lambda(n)$ is the von Mangoldt function. Hence

$$\begin{aligned}\sum_{k=1}^{\infty} \psi\left(\frac{x}{k}\right) &= \sum_{k=1}^{\infty} \sum_{n \le \frac{x}{k}} \Lambda(n) = \sum_{\substack{kn \le x \\ k \ge 1}} \Lambda(n) \\ &= \sum_{n \le x} \sum_{k \le \frac{x}{n}} \Lambda(n) = \sum_{n \le x} \left[\frac{x}{n}\right] \Lambda(n) \\ &= \sum_{n \le x} \sum_{d \mid n} \Lambda(d) = \ln[x]!,\end{aligned}$$

where the last two equalities follow from (3.8) and Exercise 2.6, Problem 1(b). □

We will now establish a few equalities and inequalities.

Lemma 4.12. *For positive real number x, we have*

$$\psi(x) - 2\psi(\sqrt{x}) = \theta(x) - \theta(\sqrt{x}) + \theta(\sqrt[3]{x}) - \cdots, \tag{4.20}$$

$$\ln[x]! - 2\ln[x/2]! = \psi(x) - \psi\left(\frac{x}{2}\right) + \psi\left(\frac{x}{3}\right) - \cdots, \tag{4.21}$$

$$\psi(x) - 2\psi\left(\sqrt{x}\right) \le \theta(x) \le \psi(x) \tag{4.22}$$

and

$$\begin{aligned}\psi(x) - \psi\left(\frac{x}{2}\right) &\le \ln[x]! - 2\ln[x/2]! \\ &\le \psi(x) - \psi\left(\frac{x}{2}\right) + \psi\left(\frac{x}{3}\right).\end{aligned} \tag{4.23}$$

***Proof of* (4.20).**

This follows directly from (4.18). More precisely,

$$\psi(x) - 2\psi(\sqrt{x}) = \sum_{k=1}^{\infty} \theta\left(\sqrt[k]{x}\right) - 2\sum_{k=1}^{\infty} \theta\left(\sqrt[2k]{x}\right).$$

□

***Proof of* (4.21).**

This follows from (4.19), namely,

$$\ln[x]! - 2\ln[x/2]! = \sum_{k=1}^{\infty} \psi\left(\frac{x}{k}\right) - 2\sum_{k=1}^{\infty} \psi\left(\frac{x}{2k}\right).$$

□

***Proof of* (4.22).**

Note that $\theta(x)$ is increasing. Hence, from (4.20),

$$\psi(x) - 2\psi(\sqrt{x}) \leq \theta(x).$$

Also, from (4.18),

$$\psi(x) \geq \theta(x).$$

□

***Proof of* (4.23).**

This follows immediately from (4.21). □

Lemma 4.13. *Let x be a real number. Then*

$$\ln[x]! - 2\ln[x/2]! > \frac{2}{3}x \quad \text{if } x > 750, \tag{4.24}$$

$$\ln[x]! - 2\ln[x/2]! < \frac{3}{4}x \quad \text{if } x > 0, \tag{4.25}$$

$$\psi(x) - \psi(\left(\frac{x}{2}\right) + \psi\left(\frac{x}{3}\right) > \frac{2}{3}x \quad \text{if } x > 750, \tag{4.26}$$

and

$$\psi(x) - \psi\left(\frac{x}{2}\right) < \frac{3}{4}x \quad \text{if } x > 0. \tag{4.27}$$

***Proof of* (4.24).**

For real number z, the Gamma function $\Gamma(z)$ is given by

$$\Gamma(z) = \lim_{n\to\infty} \frac{(n-1)!n^z}{z(z+1)\cdots(z+n-1)}.$$

The function $\Gamma(x)$ satisfies the well-known Stirling's formula [14, p. 253]

$$\ln\Gamma(x) = \ln\sqrt{2\pi} + \left(x - \frac{1}{2}\right)\ln x - x + \frac{\vartheta_x}{12x}, \qquad 0 < \vartheta_x < 1. \tag{4.28}$$

Since $\Gamma(x)$ is increasing when $x > 3$, we conclude that

$$\ln[x]! - 2\ln[x/2]! \geq \ln\Gamma(x) - 2\ln\Gamma\left(\frac{1}{2}x + 1\right).$$

To prove (4.24), it suffices to show that for $x > 750$,

$$\ln\Gamma(x) - 2\ln\Gamma\left(\frac{1}{2}x + 1\right) > \frac{2x}{3}. \tag{4.29}$$

By (4.28), we deduce that

$$\begin{aligned}
\ln\Gamma(x) - 2\ln\Gamma\left(\frac{1}{2}x+1\right) &= \ln\sqrt{2\pi} + \left(x-\frac{1}{2}\right)\ln x - x + \frac{\vartheta_1}{12x} - 2\ln\sqrt{2\pi} \\
&\quad - 2\left(\frac{x}{2}+\frac{1}{2}\right)\ln\left(\frac{x}{2}+1\right) + 2\left(\frac{x}{2}+1\right) - \frac{\vartheta_2}{3x+6},
\end{aligned}$$

where both ϑ_1, ϑ_2 belong to the interval $(0,1)$. Simplifying the above inequality, we deduce that

$$\ln\Gamma(x) - 2\ln\Gamma\left(\frac{1}{2}x+1\right) > x\ln\left(\frac{2x}{x+2}\right) - 2\ln x.$$

For $x > 750$,

$$x\ln\left(\frac{2x}{x+2}\right) - 2\ln x > \frac{2x}{3}.$$

This completes the proof of (4.29). □

***Proof of* (4.25).**

The proof is similar to that for (4.24). We use the inequality

$$\ln[x]! - 2\ln[x/2]! \le \ln\Gamma(x+1) - 2\ln\Gamma\left(\frac{1}{2}x+\frac{1}{2}\right)$$

and Stirling's formula to conclude that (see Problem 8 of Exercise 4.7)

$$\ln[x]! - 2\ln[x/2]! \le \frac{3}{4}x$$

for all $x > 0$. □

***Proof of* (4.26) *and* (4.27).**

These two inequalities follow immediately from (4.23)–(4.25). □

Lemma 4.14. *For each positive real number x, we have*

$$\psi(x) < \frac{3}{2}x \quad \textit{if } x > 0 \tag{4.30}$$

$$\begin{aligned}
\psi(x) - \psi\left(\frac{x}{2}\right) + \psi\left(\frac{x}{3}\right) &\le \theta(x) + 2\psi(\sqrt{x}) - \theta\left(\frac{x}{2}\right) + \psi\left(\frac{x}{3}\right) \\
&< \theta(x) - \theta\left(\frac{x}{2}\right) + \frac{x}{2} + 3\sqrt{x}\,.
\end{aligned} \tag{4.31}$$

***Proof of* (4.30).**

To prove (4.30), we use (4.27) repeatedly, with x replaced by $x/2$, $x/4, \cdots$ and add up the results. We find that

$$\psi(x) \leq \frac{3}{4}x\left(1+\frac{1}{2}+\cdots\right) < \frac{3}{2}x.$$

***Proof of* (4.31).**

From (4.22), we find that

$$\psi(x) - 2\psi(\sqrt{x}) \leq \theta(x).$$

Hence

$$\psi(x) \leq \theta(x) + 2\psi(\sqrt{x}).$$

Next, from (4.22),

$$\theta(x/2) \leq \psi(x/2).$$

Using the above inequalities, we deduce that

$$\psi(x) - \psi(x/2) + \psi(x/3) \leq \theta(x) + 2\psi(\sqrt{x}) - \theta(x/2) + \psi(x/3).$$

For the second inequality, we use (4.30) to deduce that

$$2\psi(\sqrt{x}) + \psi(x/3) \leq 3\sqrt{x} + x/2.$$

We are now ready to prove Bertrand's Postulate. By (4.26),

$$\psi(x) - \psi(x/2) + \psi(x/3) \geq \frac{2}{3}x$$

for $x > 750$. Hence we deduce from (4.31) that

$$\theta(x) - \theta(x/2) \geq 2x/3 - x/2 - 3\sqrt{x},$$

and there is a prime between x and $2x$ for $x > 750$.

We are now left with verifying that Bertrand's Postulate is true for $x \leq 750$. This is straightforward and we leave it as an exercise.

Example 4.1. Suppose S is a set of consecutive integers which contains a prime. Show that there exists an integer in S that is relatively prime to all the other integers in S. Conversely, show that if for every finite set of consecutive integers with at least a prime contains a number that is relatively prime to the others, then Bertrand's Postulate is true.

Proof. Let $S := \{n, n+1, \cdots, n+k\}$ be any set of consecutive integers with at least a prime. Suppose p is the largest prime in the set S. If $2p \leq n+k$, then by Bertrand's Postulate, there exists another prime q such that $p < q < 2p$ and this contradicts our assumption that p is the largest prime in the set. Hence $2p > n+k$ and p is relatively prime to all the other integers in S.

Conversely, let $n > 1$ be given and consider the set $T = \{2, 3, \cdots, 2n\}$. Obviously there is at least a prime in T and hence by hypothesis, there exists an integer q such that q is relatively prime to the other numbers in T. This implies that $2q > 2n$ (for, if $2q < 2n$ then q will not be prime to $2q$.) Hence $n < q < 2n$. To prove that q is prime, we suppose $a|q$ and $a \neq 1$. Then $a \in T$ and so, a is not relatively prime to q. This contradicts the choice of q. Hence q must be prime. This means that there is a prime between n and $2n$ if $n > 1$, and this is Bertrand's Postulate. □

4.7 Exercises

1. Prove that the following two relations are equivalent.

   (a) $\pi(x) = \dfrac{x}{\ln x} + O\left(\dfrac{x}{\ln^2 x}\right)$

   (b) $\theta(x) = x + O\left(\dfrac{x}{\ln x}\right)$.

2. Show that if p, q are primes, then
$$\sum_{pq \leq x} \frac{1}{pq} = (\ln \ln x)^2 + O(\ln \ln x).$$

3. Let $\omega(n)$ be defined as in Definition 3.6. Show that
$$\sum_{n \leq x} \omega(n) = x \ln \ln x + O(x).$$

4. Show that for $x \geq 2$,
$$\sum_{n \leq x} \psi\left(\frac{x}{n}\right) = x \ln x + O(x).$$

5. Show that for $x \geq 1$,

$$\sum_{n \leq x} \frac{\theta(n)}{n^2} = \ln x + O(1).$$

6. Show that

$$\sum_{p \leq x} \frac{\ln p}{p-1} = \ln x + \mathrm{O}(1).$$

7. Let x be a positive real number. Verify using Stirling's formula that

$$\ln[x]! - 2\ln[x/2]! < \frac{3}{4}x \quad \text{if } x > 0.$$

8. Follow Erdös' proof of Bertrand's Postulate by completing the following exercises.

   (a) Let $r(p)$ be such that

   $$p^{r(p)} \leq 2n < p^{r(p)+1}.$$

   Show that

   $$\binom{2n}{n} \mid \prod_{p \leq 2n} p^{r(p)}.$$

   (b) Show that if $p > 2$ and $\frac{2n}{3} < p \leq n$ then

   $$p \nmid \binom{2n}{n}.$$

   (c) Show that

   $$\prod_{p \leq n} p < 4^n.$$

   (d) Use the above results to deduce Bertrand's Postulate.

9. Show that if n is any positive integer greater than 1, then $n!$ is never a perfect square.

10. Let n be a positive integer. Assume that when $m > 20$, there exists a prime between $m/2$ and $m - 6$. Show that n can be expressed as a sum of distinct primes when $n > 6$. (Challenge: Try to prove the above result without making the first assumption.)

11. Use Bertrand's Postulate to show that for every positive integer n, the number

$$1 + \frac{1}{2} + \frac{1}{3} + \cdots + \frac{1}{n}$$

can never be an integer.

12. Use induction and Bertrand's Postulate to show that if p_n denote the n-th prime, then for $n > 3$,

$$p_n < p_1 + p_2 + \cdots + p_{n-1}.$$

Chapter 5

The Prime Number Theorem

5.1 The Prime Number Theorem

In Chapter 4, Corollary 4.6, we proved that the Prime Number Theorem is equivalent to the statement

$$\psi(x) \sim x. \tag{5.1}$$

In this chapter, we will prove the following theorem.

Theorem 5.1. *For positive real number x, we have*

$$\psi(x) = x + O\left(x \exp(-c \sqrt[10]{\ln x})\right),$$

where $c > 0$ is some constant independent of x.

We note that (5.1) follows immediately from Theorem 5.1.

Theorem 5.1, which was mentioned in [11, p. 169], is weaker than the result obtained independently by J. Hadamard and de la Valleé Poussin, which states that

$$\psi(x) = x + O\left(x \exp(-c\sqrt{\ln x})\right).$$

But the treatment here (adapted from A. Hildebrand's 1991 "Analytic Number Theory" notes [5]) allows us to appreciate the analytic method used in the proofs of the Prime Number Theorem with less technicalities.

5.2 The Riemann zeta function

In Chapter 3, Definition 3.3, we have encountered the Riemann zeta function for real $s > 1$. We now give the definition of the function when s is a complex number.

Definition 5.1. Let $s = \sigma + it \in \mathbf{C}$ and $\sigma > 1$. Define

$$\zeta(s) = \sum_{n=1}^{\infty} \frac{1}{n^s}.$$

Theorem 5.2. *The Riemann zeta function $\zeta(s)$ is an analytic function for $\sigma > 1$.*

Proof. Note that if $\sigma \geq 1 + \delta$, then

$$\sum_{n=m}^{M} \left| \frac{1}{n^s} \right| \leq \sum_{n=m}^{M} \frac{1}{n^\sigma} \leq \sum_{n=m}^{M} \frac{1}{n^{1+\delta}}.$$

Now, for every $\epsilon > 0$, there exists $N_\epsilon > 0$ such that

$$\sum_{n=m}^{M} \frac{1}{n^{1+\delta}} < \epsilon$$

for $M > m > N_\epsilon$. Hence, we conclude that

$$\sum_{n=m}^{M} \left| \frac{1}{n^s} \right| < \epsilon$$

for $M > m > N_\epsilon$. Therefore, by the Weierstrass M-test, the series

$$\sum_{n=1}^{\infty} \frac{1}{n^s}$$

is absolutely and uniformly convergent in any region $\sigma \geq 1 + \delta$, with $\delta > 0$. The Riemann zeta function $\zeta(s)$ is therefore an analytic function in $\sigma > 1$. □

5.3 Euler's product and the product representation of $\zeta(s)$

Theorem 5.3. *For $\sigma > 1$,*

$$\zeta(s) = \prod_{p} \left(1 - \frac{1}{p^s}\right)^{-1}.$$

The above follows immediately from the next theorem.

Definition 5.2. An infinite product

$$\prod_{n=1}^{\infty}(1+a_n)$$

is said to be absolutely convergent if

$$\sum_{n=1}^{\infty}\ln(1+a_n)$$

is absolutely convergent.

Theorem 5.4. *Let f be a multiplicative arithmetical function such that the series*

$$\sum_{n=1}^{\infty} f(n)$$

is absolutely convergent. Then the sum of the series can be expressed as an absolutely convergent infinite product, namely,

$$\sum_{n=1}^{\infty} f(n) = \prod_{p}(1 + f(p) + f(p^2) + \cdots), \tag{5.2}$$

extended over all primes.

The product above is called the Euler product of the series.

Proof. Consider the finite product

$$P(x) = \prod_{p\le x}(1 + f(p) + f(p^2) + \cdots)$$

extended over all primes $p \le x$. Since this is a product of a finite number of absolutely convergent series we can multiply the series and rearrange the terms without altering the sum. A typical term is of the form

$$\prod_{p} f(p^{\alpha}) = f\left(\prod_{p} p^{\alpha}\right),$$

since f is multiplicative. By the fundamental theorem of arithmetic we can write

$$P(x) = \sum_{n\in A} f(n)$$

where A consists of those n having all their prime factors less than or equal to x. Therefore,

$$\sum_{n=1}^{\infty} f(n) - P(x) = \sum_{n \in B} f(n),$$

where B is the set of n having at least one prime factor greater than x. Therefore,

$$\left| \sum_{n=1}^{\infty} f(n) - P(x) \right| \leq \sum_{n \in B} |f(n)| \leq \sum_{n > x} |f(n)|.$$

Since

$$\sum_{n=1}^{\infty} |f(n)|$$

is convergent,

$$\lim_{x \to \infty} \sum_{n > x} |f(n)| = 0.$$

Hence,

$$\lim_{x \to \infty} P(x) = \sum_{n=1}^{\infty} f(n).$$

We have proved that the infinite product is convergent. We now establish the absolute convergence of the infinite product. A necessary and sufficient condition for the absolute convergence of the product

$$\prod_{n} (1 + a_n)$$

is the convergence of the series (see [1, p. 192])

$$\sum_{n} |a_n|.$$

In this case, we have

$$\sum_{p \leq x} |f(p) + f(p^2) + f(p^3) + \cdots| \leq \sum_{p \leq x} (|f(p)| + |f(p^2)| + \cdots) \leq \sum_{n=2}^{\infty} |f(n)|.$$

Since the partial sums are bounded, the series of positive terms

$$\sum_{p \leq x} |f(p) + f(p^2) + f(p^3) + \cdots|$$

converges, and this implies absolute convergence of the product (5.2). □

Applying Theorem 5.4 with

$$f(n) = \frac{1}{n^s},$$

we obtain Theorem 5.3.

5.4 Analytic continuation of $\zeta(s)$ to $\sigma > 0$

Theorem 5.5. *The Riemann zeta function $\zeta(s)$ can be extended to a function that is analytic in $\sigma > 0$, except at $s = 1$, where it has a simple pole with residue 1.*

Proof. Recall from Theorem 3.2 that

$$\sum_{n \le x} f(n) = f(1) + \int_1^x f(t)dt + \int_1^x f'(t)\{t\}dt - \{x\}f(x).$$

With s real,

$$x = N \in \mathbf{N} \quad \text{and} \quad f(n) = \frac{1}{n^s},$$

we have

$$\sum_{n=1}^{N} \frac{1}{n^s} = 1 + \int_1^N \frac{d\eta}{\eta^s} - \int_1^N \frac{s\{\eta\}}{\eta^{s+1}} d\eta.$$

By analytic continuation, the above identity is also valid for complex numbers $s = \sigma + it$ with $\sigma > 1$.

Now, assume $\sigma > 1$. Then

$$\lim_{N \to \infty} \sum_{n=1}^{N} \frac{1}{n^s} = \zeta(s),$$

$$\lim_{N \to \infty} \int_1^N \frac{d\eta}{\eta^s} = \int_1^\infty \frac{d\eta}{\eta^s} = \frac{1}{s-1}$$

and

$$\lim_{N \to \infty} \int_1^N \frac{\{\eta\}}{\eta^{s+1}} d\eta = \int_1^\infty \frac{\{\eta\}}{\eta^{s+1}} d\eta =: \Phi(s).$$

Therefore,

$$\zeta(s) = 1 + \frac{1}{s-1} - s\Phi(s), \sigma > 1.$$

But, $\Phi(s)$ is analytic for $\sigma > 0$. Define, for $\sigma > 0$, the extension of $\zeta(s)$ by

$$1 + \frac{1}{s-1} - s\Phi(s).$$

Note that this function has a pole at $s = 1$. □

5.5 Upper bounds for $|\zeta(s)|$ and $|\zeta'(s)|$ near $\sigma = 1$

Theorem 5.6. *Let A be a positive real number. If*

$$|t| \geq 2 \quad \text{and} \quad \sigma \geq \max\left(\frac{1}{2}, 1 - \frac{A}{\ln|t|}\right), \tag{5.3}$$

then there are positive constants M and M' (depending on A) such that

$$|\zeta(s)| \leq M \ln|t| \tag{5.4}$$

$$|\zeta'(s)| \leq M' \ln^2|t|. \tag{5.5}$$

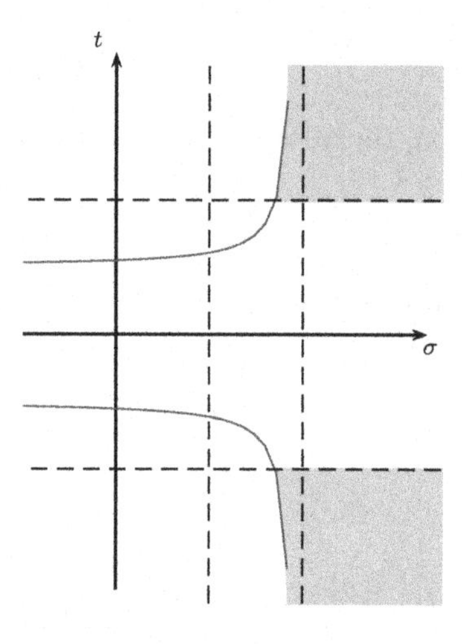

Fig. 5.1 The shaded regions indicate the regions for which (5.4) and (5.5) hold.

Proof. Suppose s is real and $s > 1$. Then by the Euler-Maclaurin summation formula (see Theorem 3.2),

$$\begin{aligned}\sum_{n=1}^{N}\frac{1}{n^s} &= 1+\int_1^N\frac{dx}{x^s}-s\int_1^N\frac{\{x\}}{x^{s+1}}dx\\ &= 1+\frac{N^{1-s}-1}{1-s}-s\Phi(s)+s\int_N^\infty\frac{\{x\}}{x^{s+1}}dx\\ &= \zeta(s)+\frac{N^{1-s}}{1-s}+s\int_N^\infty\frac{\{x\}}{x^{s+1}}dx.\end{aligned}$$

By analytic continuation, the above identity holds for complex number s with $\sigma > 1$.

Now,

$$\begin{aligned}|\zeta(s)| &\le \sum_{n=1}^{N}\frac{1}{n^\sigma}+\frac{N^{1-\sigma}}{|1-s|}+|s|\int_N^\infty\frac{dx}{x^{\sigma+1}}\\ &\le \sum_{n=1}^{N}\frac{1}{n^\sigma}+\frac{N^{1-\sigma}}{|t|}+\frac{|s|}{\sigma N^\sigma}.\end{aligned}$$

Assume that s is in the region specified by (5.3) and let $N = [|t|]$. Then

$$N^{1-\sigma} \le \exp\left\{\frac{A\ln N}{\ln|t|}\right\} \le \exp(A).$$

This implies that

$$\begin{aligned}|\zeta(s)| &\le \sum_{n\le|t|}\frac{1}{n^\sigma}+\frac{e^A}{|t|}+\frac{|s|}{\sigma N}e^A\\ &\le \sum_{n\le|t|}\frac{1}{n^\sigma}+\frac{e^A}{2}+\frac{(\sigma+|t|)e^A}{\sigma(|t|/2)}\\ &\le \sum_{n\le|t|}\frac{1}{n^\sigma}+e^A\left(\frac{1}{2}+\frac{2}{|t|}+\frac{2}{\sigma}\right).\end{aligned} \tag{5.6}$$

Since $\sigma > 1/2$ and $|t| \ge 2$, we find that

$$\frac{1}{2}+\frac{2}{|t|}+\frac{2}{\sigma} < 6.$$

This shows that (5.6) may be written as

$$|\zeta(s)| \le \sum_{n\le|t|}\frac{1}{n^\sigma}+6e^A. \tag{5.7}$$

For $\sigma \ge 1$,

$$\sum_{n\le|t|}\frac{1}{n^\sigma} \le \sum_{n\le|t|}\frac{1}{n} = \ln|t| + O(1). \tag{5.8}$$

For

$$\max\left(\frac{1}{2}, 1 - \frac{A}{\ln|t|}\right) < \sigma < 1,$$

and $n \leq N$, we find that

$$\frac{1}{n^\sigma} \leq \frac{1}{n} n^{1-\sigma} \leq \frac{1}{n} N^{1-\sigma} \leq \frac{1}{n}\left(1 + e^A - 1\right).$$

Hence,

$$\sum_{n\leq|t|} \frac{1}{n^\sigma} \leq e^A \sum_{n\leq|t|} \frac{1}{n} = e^A \left(\ln|t| + O(1)\right). \tag{5.9}$$

Combining (5.8) and (5.9), we conclude that if s is in the region specified by (5.3), then

$$|\zeta(s)| \leq M \ln|t|,$$

where M is a positive constant depending on A. This proves (5.4).

We leave the proof of (5.5) as an exercise (Problem 1 of Exercise 5.10).

□

5.6 The non-vanishing of $\zeta(1+it)$

Theorem 5.7. *For real number $t \neq 0$,*

$$\zeta(1+it) \neq 0.$$

We first prove several simple lemmas.

Lemma 5.8. *For all $\theta \in \mathbf{R}$,*

$$3 + 4\cos\theta + \cos 2\theta \geq 0.$$

Proof. The inequality follows immediately from the following computations:

$$\begin{aligned} 3 + 4\cos\theta + 2\cos^2\theta - 1 &= 2\cos^2\theta + 4\cos\theta + 2 \\ &= 2(\cos^2\theta + 2\cos\theta + 1) \\ &= 2(\cos\theta + 1)^2 \geq 0. \end{aligned}$$

□

Lemma 5.9. *For* $\sigma > 1$,

$$\zeta(s) = e^{G(s)},$$

where

$$G(s) = \sum_p \sum_{m=1}^{\infty} \frac{1}{mp^{ms}}.$$

Proof. Using the Euler product representation of $\zeta(s)$, we find that

$$\begin{aligned}
\zeta(s) &= \prod_p \left(1 - \frac{1}{p^s}\right)^{-1} \\
&= \exp\left(-\sum_p \ln\left(1 - \frac{1}{p^s}\right)\right) \\
&= \exp\left(\sum_p \sum_{m=1}^{\infty} \frac{1}{mp^{sm}}\right) = \exp(G(s)).
\end{aligned}$$

□

Lemma 5.10. *For* $\sigma > 1$, *and all* $t \in \mathbf{R}$,

$$|\zeta(\sigma)|^3 |\zeta(\sigma + it)|^4 |\zeta(\sigma + 2it)| \geq 1.$$

Proof. By Lemma 5.9, we have for $\sigma > 1$,

$$\begin{aligned}
\zeta(s) &= \exp\left(\sum_p \sum_{m=1}^{\infty} \frac{1}{mp^{ms}}\right) \\
&= \exp\left(\sum_p \sum_{m=1}^{\infty} \frac{1}{m} \exp\{-(\ln p)ms\}\right) \\
&= \exp\left(\sum_p \sum_{m=1}^{\infty} \frac{1}{m} \exp\{-m\sigma \ln p - itm \ln p\}\right),
\end{aligned}$$

since $s = \sigma + it$. Hence,

$$\zeta(s) = \exp\left(\sum_p \sum_{m=1}^{\infty} \frac{1}{m} \frac{1}{p^{\sigma m}} \{\cos(tm \ln p) - i \sin(tm \ln p)\}\right).$$

Therefore,

$$|\zeta(s)| = \exp\left(\sum_p \sum_{m=1}^{\infty} \frac{1}{m} \frac{1}{p^{\sigma m}} \cos(tm \ln p)\right).$$

This implies that

$$
\begin{aligned}
&|\zeta(\sigma))|^3|\zeta(\sigma+it)|^4|\zeta(\sigma+2it)| \\
&\qquad = \exp\left(\sum_p \sum_{m=1}^{\infty} \frac{1}{mp^{\sigma m}}(3+4\cos(tm\ln p)+\cos(2tm\ln p))\right) \\
&\qquad \geq \exp(0) = 1.
\end{aligned}
$$

□

Proof of Theorem 5.7.

Suppose $\zeta(1+it_0)=0$ for some $t_0 \neq 0$. By Lemma 5.10, we deduce that for $\sigma > 1$,

$$
|\zeta(\sigma)(\sigma-1)|^3 \left|\frac{\zeta(\sigma+it_0)}{\sigma-1}\right|^4 |\zeta(\sigma+2it_0)| \geq \frac{1}{\sigma-1}. \tag{5.10}
$$

Now, since $\zeta(\sigma)$ has a simple pole with residue 1 at $\sigma = 1$, we find that

$$
\lim_{\sigma\to 1^+} \zeta(\sigma)(\sigma-1) = 1. \tag{5.11}
$$

Next,

$$
\zeta(\sigma+it_0) = \zeta(1+it_0) + (\sigma-1)\zeta'(1+it_0) + O((\sigma-1)^2).
$$

This implies that

$$
\lim_{\sigma\to 1^+} \frac{\zeta(\sigma+it_0)}{\sigma-1} = \zeta'(1+it_0). \tag{5.12}
$$

It is clear that

$$
\zeta(\sigma+2it_0) \to \zeta(1+2it_0). \tag{5.13}
$$

Combining (5.11)–(5.13), we find that when σ approaches 1 from the right, the right-hand side of (5.10) approaches ∞ and the left-hand side of (5.10) is finite. This leads to a contradiction and we conclude that

$$
\zeta(1+it) \neq 0,
$$

for all nonzero real t. □

5.7 A lower bound for $|\zeta(s)|$ near $\sigma = 1$

Theorem 5.11. *For $|t| \geq 2$, there exist positive constants c and d such that for*

$$\sigma > 1 - \frac{c}{(\ln |t|)^9},$$

we have

$$|\zeta(\sigma + it)| \geq \frac{d}{(\ln |t|)^7}.$$

Proof. For $\sigma \geq 2$,

$$\begin{aligned} |\zeta(s)| &= \left| \sum_{n=1}^{\infty} \frac{1}{n^s} \right| \geq 1 - \left| \sum_{n=2}^{\infty} \frac{1}{n^s} \right| \\ &\geq 1 - \sum_{n=2}^{\infty} \frac{1}{n^2} = 2 - \frac{\pi^2}{6} > \frac{1}{4}. \end{aligned}$$

Therefore, for $\sigma \geq 2$,

$$|\zeta(s)| \geq \frac{d}{(\ln |t|)^7},$$

provided that

$$d \leq \frac{\ln^7(2)}{4} \quad \text{and} \quad |t| \geq 2. \tag{5.14}$$

For $\delta > 0$, let

$$1 + \frac{\delta}{(\ln |t|)^9} \leq \sigma \leq 2, \quad |t| \geq 2.$$

By Lemma 5.10, we find that

$$|\zeta(\sigma + it)| \geq \frac{1}{|\zeta(\sigma)|^{3/4} |\zeta(\sigma + 2it)|^{1/4}}.$$

Now, if $\sigma \leq 2$,

$$\begin{aligned} \zeta(\sigma) = \sum_{n=1}^{\infty} \frac{1}{n^\sigma} &\leq 1 + \int_1^{\infty} \frac{1}{x^\sigma} dx = 1 + \frac{1}{\sigma - 1} \\ &\leq \frac{2}{\sigma - 1} \\ &\leq \frac{2}{\delta} (\ln |t|)^9, \end{aligned}$$

since

$$\sigma \geq 1 + \frac{\delta}{(\ln |t|)^9}.$$

Now suppose

$$\delta < \frac{\ln^9 2}{2}. \tag{5.15}$$

Then since $|t| > 2$ we have

$$\frac{1}{\ln |t|} < \frac{1}{\ln 2}$$

and therefore,

$$\frac{\delta}{\ln^9 |t|} \leq \frac{1}{2} \frac{\ln^9 2}{\ln^8 |t| \ln |t|} < \frac{1}{2} \frac{\ln 2}{\ln |t|}.$$

In other words, if δ satisfies (5.15) and

$$\sigma > 1 - \frac{\delta}{\ln^9 |t|},$$

we must have

$$\sigma > 1 - \frac{1}{2} \frac{\ln 2}{\ln |t|}. \tag{5.16}$$

By Theorem 5.6 with $A = \frac{1}{2} \ln 2$, we can find a constant $M > 0$ such that

$$|\zeta(\sigma + 2it)| \leq 2M \ln |t|.$$

Hence,

$$\begin{aligned} |\zeta(\sigma + it)| &\geq \left(\frac{\delta}{2 \ln^9 |t|} \right)^{3/4} \left(\frac{1}{2M \ln |t|} \right)^{1/4} \\ &= \frac{\delta^{3/4}}{2M^{1/4} \ln^7 |t|} \geq \frac{d}{\ln^7 |t|}, \end{aligned}$$

for

$$d \leq \frac{\delta^{3/4}}{2M^{1/4}}. \tag{5.17}$$

Next, consider

$$1 - \frac{\delta}{\ln^9 |t|} \leq \sigma \leq 1 + \frac{\delta}{\ln^9 |t|}, \quad |t| \geq 2.$$

If

$$\sigma_0 = 1 + \frac{\delta}{\ln^9 |t|},$$

then we want to show that $\zeta(\sigma + it)$ is close to $\zeta(\sigma_0 + it)$.

$$\begin{aligned}|\zeta(\sigma + it) - \zeta(\sigma_0 + it)| &= \left| \int_{\sigma}^{\sigma_0} \zeta'(u + it) du \right| \\ &\leq |\sigma - \sigma_0| \max_{\sigma \leq u \leq \sigma_0} |\zeta'(u + it)|.\end{aligned}$$

Now, by Theorem 5.6, there exists an $M' > 0$ such that

$$|\zeta'(u + it)| \leq M' \ln^2 |t|,$$

for

$$|u| \geq \sigma \geq 1 - \frac{1}{2} \frac{\ln 2}{\ln |t|} \quad \text{and} \quad |t| \geq 2.$$

Therefore,

$$|\zeta(\sigma + it) - \zeta(\sigma_0 + it)| \leq \frac{2\delta}{\ln^7 |t|} M'.$$

Hence,

$$\begin{aligned}|\zeta(\sigma + it)| &\geq |\zeta(\sigma_0 + it)| - |\zeta(\sigma + it) - \zeta(\sigma_0 + it)| \\ &\geq \frac{\delta^{3/4}}{2M^{1/4} \ln^7 |t|} - \frac{2\delta}{\ln^7 |t|} M' \\ &= \frac{\delta^{3/4}}{\ln^7 |t|} \left(\frac{1}{2M^{1/4}} - 2\delta^{1/4} M' \right).\end{aligned}$$

We now choose a real positive number $\delta = \delta_1$ be such that

$$\left(\frac{1}{2M^{1/4}} - 2\delta_1^{1/4} M' \right) > 0.$$

Now, letting $0 < c < \min\left(\frac{1}{2} \ln^9(2), \delta_1\right)$ and

$$0 < d < \min\left(\delta_1^{3/4} \left(\frac{1}{2M^{1/4}} - 2\delta_1^{1/4} M' \right), \frac{\ln^7(2)}{4}, \frac{\delta_1^{3/4}}{2M^{1/4}} \right),$$

we complete the proof of the lemma. □

5.8 Perron's Formula

Theorem 5.12. *Let x be a half integer. Then for any $b \in [1,3]$ and any $T \geq 1$,*

$$\psi(x) = \frac{1}{2\pi i}\int_{b-iT}^{b+iT}\left(-\frac{\zeta'}{\zeta}(s)\frac{x^s}{s}\right)ds + O\left(\frac{x^b}{T(b-1)} + x\frac{\ln^2 x}{T}\right).$$

We first begin with several lemmas.

Lemma 5.13. *For $\sigma > 1$,*

$$\sum_{n=1}^{\infty}\frac{\Lambda(n)}{n^s} = -\frac{\zeta'}{\zeta}(s).$$

Proof. The proof is immediate using the formula

$$\Lambda = \mu * \ln$$

and the fact that (see Chapter 6, Theorem 6.3)

$$\sum_{n=1}^{\infty}\frac{f*g(n)}{n^s} = \sum_{n=1}^{\infty}\frac{f(n)}{n^s}\sum_{n=1}^{\infty}\frac{g(n)}{n^s}.$$

□

Lemma 5.14. *For $\sigma > 1$,*

$$\left|\frac{\zeta'}{\zeta}(s)\right| \ll \frac{1}{\sigma-1}+1.$$

Proof. For $1 < \sigma \leq 2$,

$$\begin{aligned}\left|\frac{\zeta'}{\zeta}(s)\right| &\leq \sum_{n=1}^{\infty}\frac{\Lambda(n)}{n^\sigma} = \sigma\int_1^\infty \sum_{n\leq t}\Lambda(n)\frac{dt}{t^{\sigma+1}}\\ &\leq \sigma\int_1^\infty \frac{ct}{t^{\sigma+1}}dt, \quad \text{by Theorem 4.2,}\\ &= c\frac{\sigma}{\sigma-1} \ll 1+\frac{1}{\sigma-1}.\end{aligned}$$

□

Lemma 5.15. *For $b > 0$, $T \geq 1$, and $y > 0$, $y \neq 1$, we have*

$$\frac{1}{2\pi i}\int_{b-iT}^{b+iT}\frac{y^s}{s}ds = \begin{cases} 1 + O\left(\dfrac{y^b}{T|\ln y|}\right) & \text{if } y > 1\\ O\left(\dfrac{y^b}{T|\ln y|}\right) & \text{if } 0 < y < 1\end{cases}$$

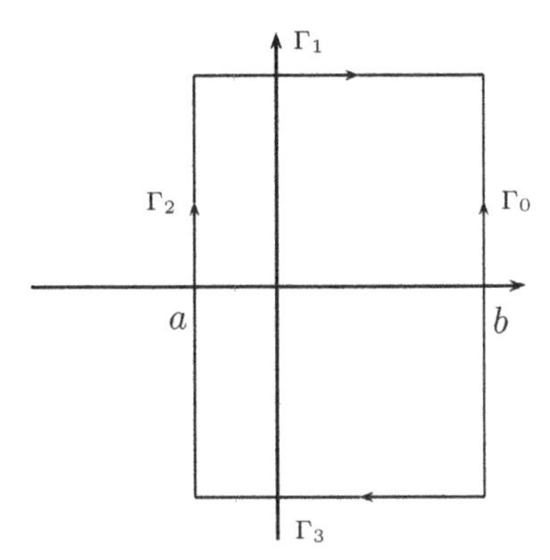

Fig. 5.2 Contours used in the proof of Lemma 5.15.

Proof. We will only prove the result when $y > 1$. By the Residue Theorem,

$$\frac{1}{2\pi i}\int_{b-iT}^{b+iT} \frac{y^s}{s}ds = \frac{1}{2\pi i}\int_{\Gamma_0} \frac{y^s}{s}ds = 1 + \frac{1}{2\pi i}\sum_{j=1}^{3}\int_{\Gamma_j} \frac{y^s}{s}ds.$$

Thus, it suffices to show that with $-a$ large enough,

$$\left|\int_{\Gamma_j} \frac{y^s}{s}ds\right| \ll \frac{y^b}{T|\ln y|},$$

with $j = 1, 2, 3$.

On Γ_2,

$$\left|\frac{y^s}{s}\right| = \frac{y^a}{|s|} \le y^a,$$

if $a \le -1$. This implies that

$$\left|\int_{\Gamma_2} \frac{y^s}{s}ds\right| \le y^a 2T.$$

Letting a approaches $-\infty$, we conclude that the above integral is 0.

On Γ_1 and Γ_3,

$$\left|\frac{y^s}{s}\right| = \frac{y^\sigma}{|s|} \le \frac{y^\sigma}{T},$$

since

$$|s| > |T|.$$

Hence, for $j = 1$ or 3,

$$\left|\int_{\Gamma_j} \frac{y^s}{s} ds\right| \leq \int_a^b \frac{y^\sigma}{T} d\sigma \leq \frac{1}{T}\int_{-\infty}^b e^{\sigma \ln y} d\sigma \ll \frac{y^b}{T|\ln y|}.$$

For the case $0 < y < 1$, we will leave it as exercise for the reader (Problem 3 of Exercise 5.10). □

Proof of Theorem 5.12.

Let

$$I = \frac{1}{2\pi i}\int_{b-iT}^{b+iT} \left(-\frac{\zeta'}{\zeta}(s)\right)\frac{x^s}{s} ds. \tag{5.18}$$

By Lemmas 5.13 and 5.15, we find that

$$\begin{aligned} I &= \frac{1}{2\pi i}\int_{b-iT}^{b+iT} \sum_{n=1}^{\infty} \frac{\Lambda(n)}{n^s}\frac{x^s}{s} ds \\ &= \sum_{n=1}^{\infty} \Lambda(n)\frac{1}{2\pi i}\int_{b-iT}^{b+iT} \frac{\left(\frac{x}{n}\right)^s}{s} ds \\ &= \sum_{n\leq x} \Lambda(n) + \sum_{n=1}^{\infty} \Lambda(n) O\left(\frac{\left(\frac{x}{n}\right)^b}{T|\ln \frac{x}{n}|}\right), \\ &= \psi(x) + O\left(\frac{x^b}{T}\sum_{n=1}^{\infty}\frac{\Lambda(n)}{n^b|\ln \frac{x}{n}|}\right). \end{aligned}$$

Let

$$R = \sum_{n=1}^{\infty} \frac{\Lambda(n)}{n^b|\ln \frac{x}{n}|}.$$

Then

$$R = \sum_{\frac{x}{2}\leq n\leq 2x} \frac{\Lambda(n)}{n^b|\ln \frac{x}{n}|} + \sum_{n\notin[\frac{x}{2},2x]} \frac{\Lambda(n)}{n^b|\ln \frac{x}{n}|} = R_1 + R_2.$$

Note that if $n > 2x$ or $n < x/2$ then $|\ln(x/n)| \geq \ln 2$. Furthermore, since $1 < b \leq 3$, by Lemmas 5.13 and 5.14,

$$R_2 \leq \frac{1}{\ln 2}\sum_{n=1}^{\infty} \frac{\Lambda(n)}{n^b} \leq 1 + \frac{1}{b-1} \leq \frac{1}{b-1} + \frac{2}{3-1} \ll \frac{1}{b-1}.$$

Now, if

$$-\frac{1}{2} \le t < 1,$$

then

$$|\ln(1+t)| \ge \frac{|t|}{2}$$

and we deduce that

$$\left|\ln\frac{x}{n}\right| = \left|\ln\left(1+\frac{x-n}{n}\right)\right| \gg \left|\frac{x-n}{n}\right|. \tag{5.19}$$

Furthermore, since

$$\Lambda(n) \le \ln x \tag{5.20}$$

and

$$\frac{1}{n^b} \le \frac{2^b}{x^b} \tag{5.21}$$

for $x/2 < n$. Using (5.19)–(5.21), we find that

$$R_1 = \sum_{\frac{x}{2}\le n\le 2x} \frac{\Lambda(n)}{n^b|\ln\frac{x}{n}|} \ll \frac{\ln x}{x^b}\sum_{\frac{x}{2}\le n\le 2x}\left|\frac{x}{x-n}\right|. \tag{5.22}$$

Since x is a half integer, the denominator in the sum

$$\sum_{\frac{x}{2}\le n\le 2x}\left|\frac{x}{x-n}\right|$$

is nonzero and we find that

$$\sum_{\frac{x}{2}\le n\le 2x}\left|\frac{x}{x-n}\right| \ll x\ln x. \tag{5.23}$$

Substituting (5.23) into (5.22), we conclude that

$$R_1 \ll 2^b\frac{\ln^2 x}{x^b}x \ll \frac{\ln^2 x}{x^b}x,$$

where we have used the fact that

$$2^b \le 2^3$$

for

$$b \le 3.$$

Hence, the error term for I, given by (5.18), is

$$O\left(\frac{x^b}{T(b-1)} + x\frac{\ln^2 T}{T}\right).$$

□

5.9 Completion of the proof of the Prime Number Theorem

Step 1.

Application of Perron's Formula:
Let

$$T \geq 1, \; x = N + \frac{1}{2} \geq 2 \quad \text{and} \quad b = 1 + \frac{1}{\ln x}.$$

Then

$$\psi(x) = \frac{1}{2\pi i} \int_{b-iT}^{b+iT} \left(-\frac{\zeta'}{\zeta}(s) \right) \frac{x^s}{s} ds + O\left(\frac{x \ln^2 x}{T} \right).$$

Step 2.

Shifting of path of integration:
Choose a sufficiently close to 1 so that

$$\zeta(s) \neq 0$$

for all $\sigma \geq a$, $|t| \leq T$. We note that the integrand

$$-\frac{\zeta'}{\zeta}(s)\frac{x^s}{s}$$

is analytic in the region enclosed by the old and new paths with an exception of a pole at $s = 1$, with residue x. By the Residue Theorem,

$$\frac{1}{2\pi i} \int_{b-iT}^{b+iT} \left(-\frac{\zeta'}{\zeta}(s) \right) \frac{x^s}{s} ds = x + \sum_{j=1}^{3} \frac{1}{2\pi i} \int_{\Gamma_j} \left(-\frac{\zeta'}{\zeta}(s) \right) \frac{x^s}{s} ds.$$

Step 3.

Estimation of $\displaystyle\int_{\Gamma_j} \left(-\frac{\zeta'}{\zeta}(s) \right) \frac{x^s}{s} ds$:
Let

$$B = \max_{s \in \Gamma_1, \Gamma_2, \Gamma_3} \left| \frac{\zeta'}{\zeta}(s) \right|.$$

The number B depends on T and will be estimated in **Step 4**.

Now, for $T \geq 2$,

$$\begin{aligned} \left| \int_{\Gamma_2} \left(-\frac{\zeta'}{\zeta}(s) \right) \frac{x^s}{s} ds \right| &\leq x^a B \int_{a-iT}^{a+iT} \frac{|ds|}{|s|} \\ &= 2x^a B \int_0^T \frac{dt}{|a+it|} \\ &\ll B x^a \ln T. \end{aligned} \tag{5.24}$$

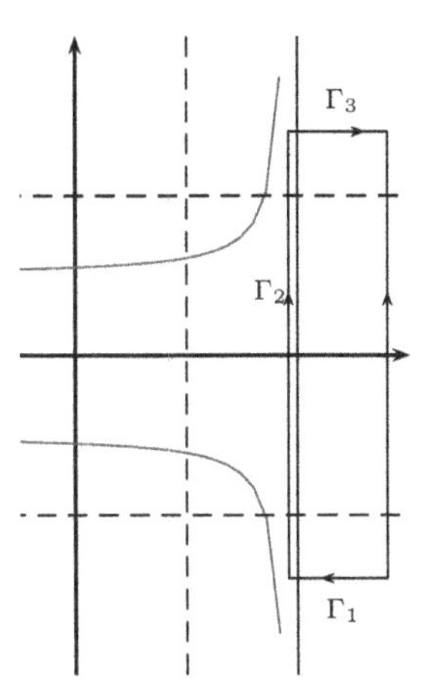

Fig. 5.3 Contours used in the proof of the Prime Number Theorem.

The last inequality follows from the fact that for $T \geq 2$,

$$\int_0^T \frac{dt}{|a+it|} \leq \int_1^T \frac{dt}{t} + \int_0^1 \frac{dt}{a} \leq \ln T + 2 \ll \ln T.$$

We will now estimate the integral on Γ_3. The estimate of the integral on Γ_1 is similar. Since

$$b = 1 + \frac{1}{\ln x},$$

we find that

$$\begin{aligned} \left| \int_{\Gamma_3} \left(-\frac{\zeta'}{\zeta}(s) \right) \frac{x^s}{s} \, ds \right| &\leq B \int_{a+iT}^{b+iT} \left| \frac{x^s}{s} \right| ds \leq \frac{B}{T} \int_a^b x^\sigma \, d\sigma \\ &\ll \frac{Bx^b}{T} \ll \frac{Bx}{T}. \end{aligned} \tag{5.25}$$

We therefore conclude from (5.24) and (5.25) that

$$\psi(x) = x + O\left(\frac{Bx}{T}\right) + O(Bx^a \ln T) + O\left(\frac{x \ln^2 x}{T}\right).$$

We note that the above holds for $T \geq 2$ and a suitable choice of a.

Step 4.

Choice of a and estimation of B:
For $|t| \leq 2$, we note that $\zeta(s) \neq 0$ for $s = 1 + it$. Therefore, there exists a $\delta > 0$ such that for $|t| \leq 2$ and $\sigma > 1 - \delta$,

$$\frac{1}{\zeta(s)}$$

is analytic and bounded there. This implies that

$$\left|\frac{\zeta'}{\zeta}(s)\right| \ll 1 \ll \ln^9 T. \tag{5.26}$$

Suppose $2 \leq |t| \leq T$. Then by Theorems 5.6 and 5.7, there exist c and d such that

$$|\zeta(s)| \geq \frac{d}{\ln^7 |t|} \quad \text{and} \quad |\zeta'(s)| \ll \ln^2 |t|$$

in the region

$$\sigma \geq 1 - \frac{c}{\ln^9 |t|}.$$

Note that we must choose c so that

$$c < \delta \ln^9 2. \tag{5.27}$$

The additional condition imposed on c is necessary for the validity of (5.26). Next, with $2 \leq |t| \leq T$, and

$$a = 1 - \frac{c}{\ln^9 |T|},$$

we conclude that

$$\left|\frac{\zeta'}{\zeta}(s)\right| \ll 1 \ll \ln^9 T.$$

Together with (5.26), we find that

$$B = \max_{s \in \Gamma_1, \Gamma_2, \Gamma_3} \left|\frac{\zeta'}{\zeta}(s)\right| \ll 1 \ll \ln^9 T.$$

Therefore,

$$\psi(x) = x + O\left(x \frac{\ln^9 T}{T}\right) + O\left(\frac{x \ln^2 x}{T}\right) + O\left(x \ln^{10} T \exp\left(-c \frac{\ln x}{\ln^9 T}\right)\right).$$

Now, the first two error terms can be bounded by

$$O\left(x\frac{\ln^{10} x}{T}\right).$$

Hence

$$\psi(x) = x + O\left(x\frac{\ln^{10} x}{T}\right) + O\left(x\ln^{10} T \exp\left(-c\frac{\ln x}{\ln^9 T}\right)\right).$$

Step 5.

Choice of T:
Assume $2 \le T \le x$. The expression in the error term is minimal if

$$\frac{1}{T} = \exp\left\{-\frac{c\ln x}{\ln^9 T}\right\}.$$

Therefore,

$$T = \exp\{c_3^{1/10}\ln^{1/10} x\}.$$

With the choice of T, we have for sufficiently large $x \ge x_0$, $2 \le T \le x$,

$$\psi(x) = x + O\left(x\frac{\ln^{10} x}{\exp(c^{1/10}\ln^{1/10} x)}\right).$$

Since for any $\epsilon > 0$,

$$\ln^{10} x \ll \exp\left(\epsilon \ln^{1/10} x\right),$$

we conclude that

$$\psi(x) = x + O(x\exp(-c'\ln^{1/10} x))$$

with a suitable choice of $c' > 0$. For $2 \le x \le x_0$, we have

$$\psi(x) = x + O(x\exp(-c'\ln^{1/10} x)).$$

This completes the proof of the Prime Number Theorem.

The equivalent statements of the above for $\theta(x)$ and $\pi(x)$ are

$$\theta(x) = x + O(x\exp(-c\ln^{1/10} x)), \quad (x \ge 2),$$

and

$$\pi(x) = Li(x) + O(x\exp(-c\ln^{1/10} x)), \quad (x \ge 2),$$

where

$$Li(x) = \int_2^x \frac{dt}{\ln t}.$$

5.10 Exercises

1. Given $A > 0$ show that there is a constant M such that for $|t| \geq 2$, $\sigma \geq \max\left(\frac{1}{2}, 1 - \frac{A}{\ln|t|}\right)$,
$$|\zeta'(s)| \leq M \ln^2 |t|.$$

2. By writing $b(x) = \{x\} - \frac{1}{2}$, show that
$$\zeta(s) = \frac{s}{s-1} - \frac{1}{2} - s\int_1^\infty \frac{b(x)}{x^{s+1}} dx.$$
Deduce that the right-hand side is an analytic continuation of $\zeta(s)$ to $\sigma > -1$ and that
$$\zeta(0) = -\frac{1}{2}.$$

3. Let $b > 0$, $T \geq 1$ and $0 < y < 1$. Show that
$$\frac{1}{2\pi i}\int_{b-iT}^{b+iT} \frac{y^s}{s}\, ds = O\left(\frac{y^b}{T|\ln y|}\right).$$

4. Prove that
$$\lim_{s\to 1}(1-s)\frac{\zeta'(s)}{\zeta(s)} = 1.$$

5. Show that for $0 < \sigma < 1$,
$$-\frac{1}{1-\sigma} < \zeta(\sigma) < 0.$$

6. Show that for $\sigma > 1$,
$$-\frac{\zeta'(s)}{\zeta(s)} = s\int_1^\infty \frac{\psi(x)}{x^{s+1}}\, dx.$$

Chapter 6

Dirichlet Series

6.1 Absolute convergence of a Dirichlet series

A Dirichlet series is a series of the form

$$\sum_{n=1}^{\infty} \frac{f(n)}{n^s}, s = \sigma + it,$$

where $f(n)$ is an arithmetical function.

Note that if $\sigma \geq a$ then $|n^s| \geq n^a$. Therefore,

$$\left| \frac{f(n)}{n^s} \right| \leq \frac{|f(n)|}{n^a}.$$

Therefore, if a Dirichlet series converges absolutely for $s = a + ib$, then by the comparison test, it also converges absolutely for all s with $\sigma \geq a$. This observation implies the following theorem.

Theorem 6.1. *Suppose the series*

$$\sum_{n=1}^{\infty} \left| \frac{f(n)}{n^s} \right|$$

does not converge for all s or diverge for all s. Then there exists a real number σ_a called the abscissa of absolute convergence, such that the series

$$\sum_{n=1}^{\infty} \frac{f(n)}{n^s}$$

converges absolutely if $\sigma > \sigma_a$ but does not converge absolutely if $\sigma < \sigma_a$.

***Proof*.** Let D be the set of all reals σ such that

$$\sum_{n=1}^{\infty} \left| \frac{f(n)}{n^s} \right|$$

diverges. Then D is not empty because the series does not converge for all s. The set D is bounded above since the series does not diverge for all s. Therefore, D has a least upper bound which we call σ_a. If $\sigma < \sigma_a$, then $\sigma \in D$, otherwise σ would be an upper bound for D smaller than the least upper bound. If $\sigma > \sigma_a$, then $\sigma \notin D$ since σ_a is an upper bound for D. This proves the theorem.

6.2 The Uniqueness Theorem

Theorem 6.2. *Let*

$$F(s) = \sum_{n=1}^{\infty} \frac{f(n)}{n^s} \quad \textit{and} \quad G(s) = \sum_{n=1}^{\infty} \frac{g(n)}{n^s}$$

be absolutely convergent for $\sigma \geq \sigma_0$. If $F(s) = G(s)$ for each s in an infinite sequence $\{s_k\}$ such that $\sigma_k \to \infty$ as $k \to \infty$, then $f(n) = g(n)$ for every n.

Proof. Let $h(n) = f(n) - g(n)$ and let $H(s) = F(s) - G(s)$. Then $H(s_k) = 0$ for each k. To prove that $h(n) = 0$ for all n we assume that $h(n) \neq 0$ for some n and obtain a contradiction.

Let N be the smallest integer n for which

$$h(n) \neq 0. \tag{6.1}$$

Then

$$H(s) = \sum_{n=N}^{\infty} \frac{h(n)}{n^s} = \frac{h(N)}{N^s} + \sum_{n=N+1}^{\infty} \frac{h(n)}{n^s}.$$

Hence,

$$h(N) = N^s H(s) - N^s \sum_{n=N+1}^{\infty} \frac{h(n)}{n^s}.$$

Putting $s = s_k$, we have $H(s_k) = 0$, and hence

$$h(N) = -N^{s_k} \sum_{n=N+1}^{\infty} \frac{h(n)}{n^{s_k}}.$$

Choose k so that $\sigma_k > c$ where $c > \sigma_a$. Then

$$|h(N)| \leq N^{\sigma_k}(N+1)^{-(\sigma_k - c)} \sum_{n=N+1}^{\infty} |h(n)| n^{-c} = \left(\frac{N}{N+1}\right)^{\sigma_k} A$$

where A is independent of k. Letting $k \to \infty$, we find that

$$\left(\frac{N}{N+1}\right)^{\sigma_k} \to 0.$$

Hence, $h(N) = 0$, a contradiction to (6.1). Consequently, $h(n) = 0$ for all positive integers n.

The above result is very useful. For example let $f(n)$ be a completely multiplicative function. Suppose

$$F(s) = \sum_{n=1}^{\infty} \frac{f(n)}{n^s}$$

and

$$G(s) = \sum_{n=1}^{\infty} \frac{f^{-1}(n)}{n^s}$$

are absolutely convergent for $\sigma \geq \sigma_0$. Then we know that

$$G(s) = 1/F(s) = \prod_p \left(1 - \frac{f(p)}{p^s}\right) = \sum_{n=1}^{\infty} \frac{\mu(n)f(n)}{n^s}.$$

By Theorem 6.2, this shows that

$$f^{-1}(n) = \mu(n)f(n).$$

6.3 Multiplication of Dirichlet series

The next theorem relates products of Dirichlet series with the Dirichlet convolution of their coefficients.

Theorem 6.3. *Given two functions $F(s)$ and $G(s)$ represented by Dirichlet series*

$$F(s) = \sum_{n=1}^{\infty} \frac{f(n)}{n^s} \quad \text{for } \sigma > a,$$

and

$$G(s) = \sum_{n=1}^{\infty} \frac{g(n)}{n^s} \quad \text{for } \sigma > b.$$

Then in the half plane where both series converge absolutely, we have

$$F(s)G(s) = \sum_{n=1}^{\infty} \frac{f * g(n)}{n^s}.$$

If

$$F(s)G(s) = \sum_{n=1}^{\infty} \frac{\alpha(n)}{n^s}$$

for all s in a sequence $\{s_k\}$ such that $\sigma_k \to \infty$ as $k \to \infty$ then $\alpha = f * g$.

Proof. For any s for which both series converge absolutely, we have

$$F(s)G(s) = \sum_{n=1}^{\infty}\sum_{m=1}^{\infty} \frac{f(n)g(m)}{(nm)^s}.$$

Because of absolutely convergence, we can multiply these series together and arrange the terms in any way we please without altering the sum. Collect together those terms for which mn is constant, say $mn = k$. The possible values of k are $1, 2, \cdots$, hence,

$$F(s)G(s) = \sum_{k=1}^{\infty} \frac{\left(\sum_{mn=k} f(n)g(m)\right)}{k^s} = \sum_{k=1}^{\infty} \frac{h(k)}{k^s}$$

where

$$h(k) = \sum_{mn=k} f(n)g(m) = f * g(k).$$

This proves the first assertion. The second assertion follows from Theorem 6.2. □

6.4 Conditional convergence of Dirichlet series

Theorem 6.4. *For every Dirichlet series, there exists $\sigma_c \in [-\infty, \infty]$ such that the series converges (conditionally) for any s with $\sigma > \sigma_c$ and diverges for any s with $\sigma < \sigma_c$. Moreover,*

$$\sigma_c \le \sigma_a \le \sigma_c + 1.$$

Proof. We will show that if

$$\sum_{n=1}^{\infty} \frac{f(n)}{n^s}$$

converges for $s = s_1$, then it also converges for every s with $\sigma > \sigma_1$.

Since

$$\sum_{n=1}^{\infty} \frac{f(n)}{n^s}$$

converges at $s = s_1$, we conclude that there exists a positive integer N_0 such that

$$\left| \sum_{x<n\le y} \frac{f(n)}{n^{s_1}} \right| \le 1$$

for all $x, y > N_0$. Now, let s with $\sigma > \sigma_1$ be given and let $y \geq x \geq N_0$. Let $\epsilon > 0$ be given. Then

$$\sum_{x<n\leq y} \frac{f(n)}{n^s} = \sum_{x<n\leq y} \frac{f(n)}{n^{s_1}} n^{s_1 - s}$$
$$= \sum_{x<n\leq y} \frac{f(n)}{n^{s_1}} y^{s_1 - s} - \int_x^y \sum_{x<n\leq t} \frac{f(n)}{n^{\sigma_1}} t^{s_1 - s - 1}(s_1 - s)dt.$$

Therefore,

$$\left| \sum_{x<n\leq y} \frac{f(n)}{n^s} \right| \leq x^{\sigma_1 - \sigma} + \int_x^y |s_1 - s| t^{\sigma_1 - \sigma - 1} dt$$
$$\leq x^{\sigma_1 - \sigma} \left(1 + \frac{|s_1 - s|}{\sigma - \sigma_1} \right) \tag{6.2}$$
$$< \epsilon$$

provided that

$$x \geq \left\{ \left(1 + \frac{|s_1 - s|}{\sigma - \sigma_1} \right) \epsilon \right\}^{-1/(\sigma - \sigma_1)}.$$

We have therefore shown that for any $\epsilon > 0$ and a fixed s with $\sigma > \sigma_1$,

$$\left| \sum_{x<n\leq y} \frac{f(n)}{n^s} \right| < \epsilon$$

whenever

$$y \geq x \geq \min \left(N_0, \left\{ \left(1 + \frac{|s_1 - s|}{\sigma - \sigma_1} \right) \epsilon \right\}^{-1/(\sigma - \sigma_1)} \right).$$

This shows the convergence of the Dirichlet series at s.

Now, let

$$\sigma_c := \inf \left\{ \operatorname{Re} s : \sum_{n=1}^{\infty} \frac{f(n)}{n^s} \text{ converges} \right\}. \tag{6.3}$$

If $\sigma > \sigma_a$ then by previous argument, we conclude that $F(s)$ is convergent whenever $\sigma = \sigma_a + \delta$, $\delta > 0$. Therefore, we conclude that $\sigma_a \geq \sigma_c$.

It remains to show that $\sigma_a \leq \sigma_c + 1$. We first show that if

$$\sum_{n=1}^{\infty} \frac{f(n)}{n^s}$$

is convergent at $s = s_1$ then it is absolutely convergent at any s with $\sigma \geq \sigma_1 + 1$. The series

$$\sum_{n=1}^{\infty} \frac{f(n)}{n^{s_1}}$$

is convergent implies that $f(n)n^{-s_1} \to 0$ as $n \to \infty$, or

$$\left| \frac{f(n)}{n^{s_1}} \right| \leq C$$

for all $n \in \mathbf{N}$ and some positive constant C. Given s with $\sigma > \sigma_1 + 1$,

$$\left| \frac{f(n)}{n^s} \right| = \left| \frac{f(n)}{n^{s_1}} \right| \frac{1}{n^{\sigma - \sigma_1}} \leq \frac{C}{n^{\sigma - \sigma_1}},$$

with $\sigma - \sigma_1 > 1$. Therefore, for any positive integer m,

$$\sum_{n=1}^{m} \left| \frac{f(n)}{n^s} \right| \leq \sum_{n=1}^{m} \frac{C}{n^{\sigma - \sigma_1}} < \infty.$$

Since $\sigma - \sigma_1 > 1$, the series

$$\sum_{n=1}^{\infty} \frac{1}{n^{\sigma - \sigma_1}}$$

converges. By comparison test, we conclude that

$$\sum_{n=1}^{\infty} \frac{f(n)}{n^s}$$

is absolutely convergent.

Now, we have shown that if

$$\sum_{n=1}^{\infty} \frac{f(n)}{n^s}$$

is convergent at $\sigma > \sigma_1 + 1$, then $\sum_{n=1}^{\infty} \frac{f(n)}{n^s}$ is absolutely convergent. This says that $\sigma_1 + 1$ is an upper bound for the set of real numbers for which the Dirichlet series is absolutely convergent, hence $\sigma_1 + 1 > \sigma_a$ since σ_a is a least upper bound. But $\sigma_1 = \sigma_c + \delta$ for arbitrary $\delta > 0$, and hence

$$\sigma_c + 1 \geq \sigma_a.$$

□

6.5 Landau's Theorem for Dirichlet series

Theorem 6.5. *A Dirichlet series*

$$F(s) = \sum_{n=1}^{\infty} \frac{f(n)}{n^s}$$

is analytic in $\sigma > \sigma_c$, where σ_c is given by (6.3).

The proof of this result follows from [1, p. 176, Theorem 1], the inequality (6.2) and the fact that in a compact set, we can bound $\sigma - \sigma_1$ and $|s_1 - s|$ by quantities independent of s. For more details of the proof, see [2, Theorem 11.11].

We now come to the main theorem of this chapter.

Theorem 6.6. *Let*

$$F(s) = \sum_{n=1}^{\infty} \frac{f(n)}{n^s}$$

be a Dirichlet series with $f(n) \geq 0$ for all $n \in \mathbb{N}$ and $\sigma_c < \infty$. Then the function $F(s)$ has a singularity at $s = \sigma_c$.

Proof. Suppose $F(s)$ is analytic at σ_c. Then there exists $\delta > 0$ such that $F(s)$ is analytic in $D_1 := \{s : |s - \sigma_c| < \delta\}$. Fix a point on the real axis, say $\sigma_0 > \sigma_c$ contained in this disc and an $\epsilon > 0$ such that the whole disc $D_2 := \{s : |s - \sigma_0| < \epsilon\}$ is inside D_1 and $\sigma_c \in D_2$ (see the following diagram).

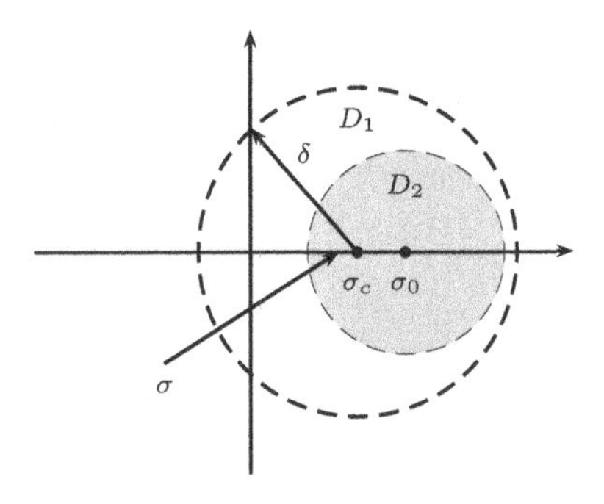

Fig. 6.1 Diagram used to illustrate the choice of discs D_1 and D_2.

Since $F(s)$ is analytic in D_2, we have the expansion

$$F(s) = \sum_{n=0}^{\infty} \frac{F^{(n)}(\sigma_0)}{n!}(s - \sigma_0)^n.$$

Now $\sigma_0 > \sigma_c$ and therefore, $F(s)$ is given by

$$F(s) = \sum_{n=1}^{\infty} \frac{f(n)}{n^s}.$$

Since $F(s)$ is analytic at σ_0, we can differentiate the above term by term to deduce that

$$F^{(\nu)}(\sigma_0) = \sum_{n=1}^{\infty} (-1)^{\nu} \frac{f(n) \ln^{\nu} n}{n^{\sigma_0}}.$$

Substituting this into the Taylor series expansion, we find that

$$F(s) = \sum_{\nu=0}^{\infty} \frac{(\sigma_0 - s)^{\nu}}{\nu!} \sum_{n=1}^{\infty} \left(\frac{f(n) \ln^{\nu} n}{n^{\sigma_0}} \right).$$

Now taking s real, say $\sigma_0 - \epsilon < s = \sigma < \sigma_0$, we have

$$\begin{aligned} F(\sigma) &= \sum_{\nu=0}^{\infty} \frac{(\sigma_0 - \sigma)^{\nu}}{\nu!} \sum_{n=1}^{\infty} \left(\frac{f(n) \ln^{\nu} n}{n^{\sigma_0}} \right) \\ &= \sum_{n=1}^{\infty} \frac{f(n)}{n^{\sigma_0}} \sum_{\nu=0}^{\infty} \frac{(\sigma_0 - \sigma)^{\nu} \ln^{\nu} n}{\nu!} \\ &= \sum_{n=1}^{\infty} \frac{f(n)}{n^{\sigma_0}} \exp((\sigma_0 - \sigma) \ln n)) \\ &= \sum_{n=1}^{\infty} \frac{f(n)}{n^{\sigma_0}} n^{\sigma_0 - \sigma} = \sum_{n=1}^{\infty} \frac{f(n)}{n^{\sigma}}. \end{aligned}$$

Hence, the Dirichlet series is convergent for some $\sigma < \sigma_c$ and this leads to a contradiction.

6.6 Exercises

1. Let $\omega(n)$ be given as in Definition 3.6. Show that

$$\sum_{n=1}^{\infty} \frac{2^{\omega(n)}}{n^s} = \frac{\zeta^2(s)}{\zeta(2s)}.$$

2. Prove that

$$\sum_{\substack{m=1 \\ (m,n)=1}}^{\infty} \sum_{n=1}^{\infty} \frac{1}{m^2 n^2} = \frac{\zeta^2(2)}{\zeta(4)}.$$

3. Show that if $\kappa(1) = 1$ and $\kappa(n) = \alpha_1\alpha_2\cdots\alpha_k$ when $n = p_1^{\alpha_1}p_2^{\alpha_2}\cdots p_k^{\alpha_k}$, then for $s > 1$,

$$\sum_{n=1}^{\infty}\frac{\kappa(n)}{n^s} = \frac{\zeta(s)\zeta(2s)\zeta(3s)}{\zeta(6s)}.$$

4. Let f be a completely multiplicative function such that $f(p) = f(p)^2$ for each prime p. If the series

$$\sum_{n=1}^{\infty}\frac{f(n)}{n^s}$$

converges absolutely for $\sigma > \sigma_a$ and has sum $F(s)$, prove that

$$\sum_{n=1}^{\infty}\frac{f(n)\lambda(n)}{n^s} = \frac{F(2s)}{F(s)}, \quad \text{if } \sigma > \sigma_a,$$

and $F(s) \neq 0$.

5. Define

$$F(\sigma,t) = 3\frac{\zeta'}{\zeta}(\sigma) + 4\frac{\zeta'}{\zeta}(\sigma+it) + \frac{\zeta'}{\zeta}(\sigma+2it).$$

If $\sigma > 1$, prove that $F(\sigma,t)$ has real part equal to

$$-\sum_{n=1}^{\infty}\frac{\Lambda(n)}{n^\sigma}\mathrm{Re}\left\{3 + 4n^{-it} + n^{-2it}\right\}$$

and deduce that

$$\mathrm{Re}\, F(\sigma,t) \leq 0.$$

6. Let $c_q(n)$ be given by Exercises 2.6, Problem 3. Show that

$$c_q(n) = \sum_{d|(n,q)} d\mu\left(\frac{q}{d}\right),$$

and deduce that

$$\frac{\sigma(n)}{n} = \frac{\pi^2}{6}\sum_{q=1}^{\infty}\frac{c_q(n)}{q^2}.$$

7. Show that for $\sigma > \sigma_a$, the abscissa of absolute convergence,

$$\sum_{n=1}^{\infty}\frac{d(n^2)}{n^s} = \frac{\zeta^3(s)}{\zeta(2s)}.$$

Hence or otherwise, show that

$$\mu^2 * d = \mu * d^2.$$

8. Show that for $\sigma > 1$,

$$\sum_{n=1}^{\infty} \sum_{m=1}^{\infty} \frac{1}{[m, n]^{\sigma}} = \frac{\zeta^3(\sigma)}{\zeta(2\sigma)}.$$

9. Show that for $\sigma > 1$,

$$\sum_{n=1}^{\infty} \frac{2^{\omega(n)} \lambda(n)}{n^{\sigma}} = \frac{\zeta(2\sigma)}{\zeta^2(\sigma)}.$$

Chapter 7

Primes in Arithmetic Progression

7.1 Introduction

In Chapter 4, we showed that there are infinitely many primes by showing that (see Theorem 4.7 (c))

$$\sum_{p \leq x} \frac{1}{p} = \ln \ln x + O(1). \tag{7.1}$$

The Dirichlet Theorem of primes in arithmetic progression states that for $(k, l) = 1$, there are infinitely many primes of the form $kn + l$. If we can prove a result similar to (7.1), with sum over primes p replaced by sum over primes of the form $kn + l$, then we would have Dirichlet's Theorem as a consequence. This strategy motivates the following theorem.

Theorem 7.1. *Let $k > 1$ and l be positive integers such that $(k, l) = 1$. Then*

$$\sum_{\substack{p \leq x \\ p \equiv l (\mathrm{mod}\ k)}} \frac{1}{p} = \frac{\ln \ln x}{\varphi(k)} + O(1).$$

Theorem 7.1 immediately implies the Dirichlet Theorem of primes in arithmetic progression.

Theorem 7.2 (Dirichlet's Theorem of primes in arithmetic progression). *If k and l are positive integers such that $(k, l) = 1$, then there are infinitely many primes of the form $kn + l$.*

7.2 Dirichlet's characters

Definition 7.1. A Dirichlet character (mod k) is an arithmetical function

$$\chi : \mathbf{N} \to \mathbf{C}$$

satisfying

(i) $\chi(mn) = \chi(m)\chi(n)$ for all $m, n \in \mathbf{N}$.

(ii) $|\chi(n)| = \begin{cases} 1 & \text{if } (n,k) = 1 \\ 0 & \text{otherwise.} \end{cases}$

(iii) $\chi(n + km) = \chi(n)$ for all $n, m \in \mathbf{N}$,

(iv) $\chi^{\varphi(k)}(n) = 1, (n,k) = 1$.

Remark 7.1.

(a) The values of χ are 0 or $\varphi(k)$-th roots of unity. This follows from (iv).

(b) There are only finitely many characters (mod k). This follows from the fact that χ is defined on $\varphi(k)$ values j with $1 \le j \le k$ and $(j,k) = 1$. Hence, from (iv), we see that for each j, there are $\varphi(k)$ values we can assign to $\chi(j)$. This shows that there can be at most $\varphi(k)^{\varphi(k)}$ characters.

(c) If χ_1 and χ_2 are characters (mod k), then so is $\chi_1\chi_2$.

(d) A character χ (mod k) can be obtained from a homomorphism

$$\widetilde{\chi} : (\mathbf{Z}/k\mathbf{Z})^* \to \{z \in \mathbf{C} \,||z| = 1\}$$

where $(\mathbf{Z}/k\mathbf{Z})^*$ is the multiplicative group of residue classes

$$(\{[n]_k | (n,k) = 1\}, \cdot),$$

with multiplication $\cdot$ as group operation. Given a character $\widetilde{\chi}$, one defines

$$\chi(n) = \begin{cases} \widetilde{\chi}([n]_k), & (n,k) = 1 \\ 0, & \text{otherwise.} \end{cases}$$

Conversely, given χ, one obtains a homomorphism $\widetilde{\chi}$ given by

$$\widetilde{\chi}([n]_k) = \chi(n).$$

This shows that there is a one to one correspondence between Dirichlet's characters (mod k) and homomorphisms from

$$(\mathbf{Z}/k\mathbf{Z})^* \text{ to } \{z \in \mathbf{C} \,||z| = 1\}.$$

Theorem 7.3. *There are exactly $\varphi(k)$ characters (mod k).*

Proof. From Remark 7.1 (d) above, it suffices to show that there are exactly $\varphi(k)$ homomorphisms from

$$(\mathbf{Z}/k\mathbf{Z})^* \text{ to } \{z \in \mathbf{C} \| z| = 1\}.$$

From the structure theorem of abelian group [8, Theorem 8.2], we know that $(\mathbf{Z}/k\mathbf{Z})^*$ can be written as a direct sums of cyclic groups with prime power order, say,

$$(\mathbf{Z}/k\mathbf{Z})^* = C_{h_1} \times \cdots \times C_{h_r},$$

where h_i are prime powers and C_m denotes a cyclic group of order m.

Let $[a_i]_k$ be a generator for C_{h_i}, $1 \le i \le r$. Given $w_1, \cdots, w_r$ such that

$$w_i^{h_i} = 1,$$

set

$$\widetilde{\chi}([a_i]_k) = w_i, \qquad 1 \le i \le r.$$

If

$$[n]_k = \prod_i [a_i]_k^{\alpha_i},$$

then define

$$\widetilde{\chi}([n]_k) = \prod_i \widetilde{\chi}([a_i]_k)^{\alpha_i}.$$

Note that $\widetilde{\chi}$ is a homomorphism from

$$(\mathbf{Z}/k\mathbf{Z})^* \text{ to } \{z \in \mathbf{C} \| z| = 1\}.$$

Therefore, we have at least

$$h_1 \cdots h_r = \varphi(k)$$

such homomorphisms.

Next, let $[a]_k \in (\mathbf{Z}/k\mathbf{Z})^*$. Then

$$[a]_k = [a_1]_k^{\alpha_1} \cdots [a_r]_k^{\alpha_r}$$

where

$$0 \le \alpha_i \le h_r - 1.$$

Now if $\widetilde{\chi}$ is a homomorphism from

$$(\mathbf{Z}/k\mathbf{Z})^* \text{ to } \{z \in \mathbf{C} \| z| = 1\},$$

then

$$\widetilde{\chi}([a]_k) = \prod_i \widetilde{\chi}([a_i]_k)^{\alpha_i}.$$

The value $\widetilde{\chi}([a]_k)$ is dependent on the values $\widetilde{\chi}([a_i]_k)$, $1 \le i \le r$. The number of possible values for $\widetilde{\chi}([a_i]_k)$ is h_i, $1 \le i \le r$. Therefore, there can be at most $h_1 h_2 \cdots h_r = \varphi(k)$ characters. In conclusion, we deduce that there are exactly $\varphi(k)$ characters (mod k).

The character χ_0 will always denote the principal character (mod k), that is,

$$\chi_0(n) = \begin{cases} 1 & \text{if } (n,k) = 1 \\ 0 & \text{otherwise.} \end{cases}$$

The character $\overline{\chi}$ will denote the inverse of χ, or, $\chi \cdot \overline{\chi} = \chi_0$.

7.3 The orthogonal relations

In this section, we will often identify (see Remark 7.1 (d)) Dirichlet's characters χ with homomorphism $\widetilde{\chi}$ from

$$(\mathbf{Z}/k\mathbf{Z})^* \quad \text{to} \quad \{z \in \mathbf{C} \,||z| = 1\}.$$

Theorem 7.4.

(a) *Let χ_1, χ_2 be two Dirichlet's characters modulo k. Then*

$$\sum_{a=1}^{k} \chi_1(a)\overline{\chi_2(a)} = \begin{cases} \varphi(k) & \textit{if } \chi_1 = \chi_2, \\ 0 & \textit{otherwise.} \end{cases}$$

(b) *Let a_1, a_2 be integers with $(a_i, k) = 1$. Then*

$$\sum_{\chi (\bmod k)} \chi(a_1)\overline{\chi(a_2)} = \begin{cases} \varphi(k) & \textit{if } a_1 \equiv a_2 \pmod{k}, \\ 0 & \textit{otherwise.} \end{cases}$$

Proof of (a).

We will prove the following:

$$\sum_{a=1}^{k} \chi(a) = \begin{cases} \varphi(k) & \text{if } \chi = \chi_0, \\ 0 & \text{otherwise.} \end{cases} \tag{7.2}$$

We first observe that since $\chi(l) = 0$ whenever $(l, k) \neq 1$, we must have

$$\sum_{a=1}^{k} \chi(a) = \sum_{\substack{a=1 \\ (a,k)=1}}^{k} \chi(a).$$

If $\chi = \chi_0$ then for $(a, k) = 1$, $\chi(a) = 1$ and

$$\sum_{a=1}^{n} \chi(a) = \sum_{\substack{a=1 \\ (a,k)=1}}^{k} \chi(a) = \sum_{\substack{a=1 \\ (a,k)=1}}^{k} 1 = \varphi(k).$$

If $\chi \neq \chi_0$, then there exists an a_0 relatively prime to k such that $\chi(a_0) \neq 1$. Now,

$$\begin{aligned}\chi(a_0) \sum_{a=1}^{k} \chi(a) &= \widetilde{\chi}([a_0]_k) \sum_{[a]_k \in (\mathbf{Z}/k\mathbf{Z})^*} \widetilde{\chi}([a]_k) \\ &= \sum_{[a]_k \in (\mathbf{Z}/k\mathbf{Z})^*} \widetilde{\chi}([a_0]_k [a]_k).\end{aligned}$$

Now, the multiplication of elements in $(\mathbf{Z}/k\mathbf{Z})^*$ by $[a_0]_k$ permutes the elements in $(\mathbf{Z}/k\mathbf{Z})^*$. Hence,

$$\sum_{[a]_k \in (\mathbf{Z}/k\mathbf{Z})^*} \widetilde{\chi}([a_0]_k [a]_k) = \sum_{[a]_k \in (\mathbf{Z}/k\mathbf{Z})^*} \widetilde{\chi}([a]_k) = \sum_{a=1}^{k} \chi(a).$$

Therefore, we conclude that

$$\sum_{a=1}^{k} \chi(a) = 0.$$

We now let $\chi = \chi_1 \overline{\chi_2}$ in (7.2) to complete the proof of (a). □

Proof of (b).

We will first show that

$$\sum_{\chi (\text{mod } k)} \chi(a) = \begin{cases} \varphi(k) & \text{if } a \equiv 1 \pmod{k}, \\ 0 & \text{otherwise.} \end{cases} \tag{7.3}$$

If $a \equiv 1 \pmod{k}$ then $\chi(a) = 1$ for all characters χ. Since there are exactly $\varphi(k)$ such characters, we conclude that

$$\sum_{\chi} \chi(a) = \varphi(k).$$

Next, suppose $a \not\equiv 1 \pmod{k}$. Then there exists a character χ^* so that $\chi^*(a) \neq 1$. Therefore,

$$\chi^*(a) \sum_{\chi} \chi(a) = \sum_{\chi} \chi^* \chi(a) = \sum_{\chi} \chi(a),$$

where we have used the fact that multiplying the elements in the set of characters by χ^* permutes the elements in the set. This implies that

$$\sum_{\chi} \chi(a) = 0.$$

Now, in order to prove (b), we simply view χ as $\widetilde{\chi}$ and let $[a]_k = [a_1]_k \overline{[a_2]}_k$ where $\overline{[a]}_k$ denotes the inverse of $[a]_k$ in the group $(\mathbf{Z}/k\mathbf{Z})^*$, and observe that

$$\chi(a_1)\overline{\chi(a_2)} = \widetilde{\chi}([a_1]_k)\widetilde{\chi}(\overline{[a_2]}_k). \qquad \square$$

7.4 The Dirichlet L-series

Definition 7.2. The Dirichlet L-series is defined as

$$L(s,\chi) = \sum_{n\geq 1} \frac{\chi(n)}{n^s}, \quad \sigma > 1.$$

Theorem 7.5.

(a) *If $\chi = \chi_0$ then $L(s,\chi)$ can be analytically continued to the half-plane $\sigma > 0$, with the exception of the point $s = 1$ where it has a simple pole with residue $\varphi(k)/k$.*
(b) *If χ is not the principal character (mod k), then $L(s,\chi)$ can be analytically continued to $\sigma > 0$.*

Proof of (a).

For $\sigma > 1$, we have by Theorem 5.4,

$$L(s,\chi) = \prod_p \left(1 - \frac{\chi(p)}{p^s}\right)^{-1}.$$

Therefore,

$$\begin{aligned} L(s,\chi_0) &= \prod_{p\nmid k} \left(1 - \frac{1}{p^s}\right)^{-1} = \prod_p \left(1 - \frac{1}{p^s}\right)^{-1} \prod_{p\mid k} \left(1 - \frac{1}{p^s}\right) \\ &= \frac{\varphi(k)}{k} \prod_p \left(1 - \frac{1}{p^s}\right). \end{aligned}$$

The function $\zeta(s)$ has an analytic continuation with residue 1 at $s = 1$. Therefore, the residue of $L(s, \chi_0)$ at $s = 1$ is $\varphi(k)/k$. □

Proof of (b).

If $\chi \neq \chi_0$, then

$$\sum_{n=1}^{k} \chi(n) = 0.$$

Therefore,

$$\left| \sum_{n \leq x} \chi(n) \right| \leq k,$$

for $x \geq 1$. Hence, for any $\epsilon > 0$,

$$\left| \sum_{y \leq n \leq x} \frac{\chi(n)}{n^s} \right| \leq \frac{1}{|y^s|} \left| \sum_{y \leq n \leq x} \chi(n) \right| < \frac{k}{|y|^\sigma} < \epsilon,$$

whenever

$$|y| > \left(\frac{k}{\epsilon}\right)^{1/\sigma}.$$

This implies that L-series converges for $\sigma > 0$. □

7.5 Proof of Dirichlet's Theorem

Step 1.

It suffices to show that if $x \geq 3$ and

$$\sigma = 1 + \frac{1}{\ln x},$$

then

$$\sum_{\substack{p \\ p \equiv l (\text{mod } k)}} \frac{1}{p^\sigma} = \frac{1}{\varphi(k)} \ln\left(\frac{1}{\sigma - 1}\right) + O(1).$$

Let

$$\Sigma_1 = \sum_{p \equiv l (\text{mod } k)} \frac{1}{p^\sigma} \quad \text{and} \quad \Sigma_2 = \sum_{\substack{p \leq x \\ p \equiv l (\text{mod } k)}} \frac{1}{p},$$

where

$$\sigma = 1 + \frac{1}{\ln x}.$$

Then

$$|\Sigma_1 - \Sigma_2| \leq \underbrace{\sum_{p \leq x} \left(\frac{1}{p} - \frac{1}{p^\sigma} \right)}_{\Sigma_3} + \underbrace{\sum_{p > x} \frac{1}{p^\sigma}}_{\Sigma_4}.$$

Now,

$$\begin{aligned}
\Sigma_3 &= \sum_{p \leq x} \frac{1 - e^{-(\sigma-1)\ln p}}{p} \leq \sum_{p \leq x} \frac{(\sigma - 1)\ln p}{p} \\
&= \frac{1}{\ln x} \sum_{p \leq x} \frac{\ln p}{p} = O(1),
\end{aligned}$$

and

$$\begin{aligned}
\Sigma_4 &= \lim_{y \to \infty} \sum_{x \leq p \leq y} \frac{1}{p^\sigma} \\
&= \lim_{y \to \infty} \left(\frac{1}{y^\sigma} \sum_{p \leq y} 1 - \frac{1}{x^\sigma} \sum_{p \leq x} 1 - \int_x^y \sum_{p \leq t} 1 \left(-\frac{\sigma}{t^{\sigma+1}} \right) dt \right) \\
&= O(1) + \int_x^\infty O\left(\frac{t}{\ln t} \right) \frac{dt}{t^{\sigma+1}} \\
&= O(1) + O\left(\int_x^\infty \frac{dt}{t^\sigma \ln t} \right) \\
&= O(1) + O\left(\frac{1}{\ln x} \int_x^\infty \frac{dt}{t^\sigma} \right) = O(1).
\end{aligned}$$

Therefore, if

$$\sigma = 1 + \frac{1}{\ln x},$$

then

$$\sum_{\substack{p \\ p \equiv l (\text{mod } k)}} \frac{1}{p^\sigma} = \frac{1}{\varphi(k)} \ln \frac{1}{\sigma - 1} + O(1)$$

and Dirichlet's Theorem holds.

Step 2.

We observe that for $\sigma > 1$,

$$\sum_{\substack{p \\ p\equiv l(\mathrm{mod}\ k)}} \frac{1}{p^\sigma} = \sum_p \frac{1}{p^\sigma}\left(\frac{1}{\varphi(k)} \sum_{\chi(\mathrm{mod}\ k)} \overline{\chi(l)}\chi(p)\right)$$
$$= \frac{1}{\varphi(k)} \sum_{\chi(\mathrm{mod}\ k)} \overline{\chi(l)} S(\sigma, \chi),$$

where

$$S(\sigma, \chi) = \sum_p \frac{\chi(p)}{p^\sigma}.$$

Now,

$$\sum_p \sum_{m\geq 1} \frac{1}{mp^{m\sigma}} - \sum_p \frac{1}{p^\sigma} = -\ln\left(1 - \frac{1}{p^\sigma}\right) - \sum_p \frac{1}{p^\sigma} = O(1),$$

since

$$\sum_p \sum_{m\geq 2} \frac{1}{mp^{m\sigma}} \leq \frac{1}{2}\sum_p \sum_{m\geq 2} \frac{1}{p^{m\sigma}} = \frac{1}{2}\sum_p \frac{1}{p^\sigma(p^\sigma - 1)} = O(1).$$

Therefore,

$$S(\sigma, \chi_0) = \sum_p \sum_{m\geq 1} \frac{1}{mp^{m\sigma}} - \sum_{p|k} \sum_{m\geq 1} \frac{1}{mp^{m\sigma}}$$
$$= -\sum_p \ln\left(1 - \frac{1}{p^\sigma}\right) + O(1)$$
$$= \ln \prod_p \left(1 - \frac{1}{p^\sigma}\right)^{-1} + O(1)$$
$$= \ln \zeta(\sigma) + O(1)$$
$$= \ln\left(\frac{1}{\sigma - 1}\right) + O(1).$$

The last equality follows from the fact that for σ near 1, we have

$$\zeta(\sigma) = \frac{1}{\sigma - 1} + g(\sigma),$$

where $g(\sigma)$ is a function analytic at 1. We conclude that the main term arises from the principal character χ_0. Hence, it remains to show that

$$S(\sigma, \chi) = O(1)$$

for $\sigma > 1$ and all non-principal characters $\chi \pmod{k}$.

Step 3.

Now, using computations similar to Step 2, we find that

$$\begin{aligned} S(\sigma, \chi) &= \sum_{p} \frac{\chi(p)}{p^{\sigma}} = \sum_{p} \sum_{m \geq 1} \frac{\chi(p)^m}{mp^{m\sigma}} + O(1) \\ &= -\sum_{p} \ln \left(1 - \frac{\chi(p)}{p^{\sigma}}\right)^{-1} + O(1) \\ &= \ln(L(\sigma, \chi)) + O(1). \end{aligned}$$

Now, for $\chi \neq \chi_0$ $L(s, \chi)$ is analytic in $\sigma > 0$. So, $L(\sigma, \chi)$ is continuous at $\sigma > 1$ and

$$\lim_{\sigma \to 1} L(\sigma, \chi) = L(1, \chi).$$

If $L(1, \chi) \neq 0$, then we are done. It remains to show that $L(1, \chi) \neq 0$.

Step 4.

We first show that when $\chi \neq \chi_0$ is a complex character (mod k), then

$$L(1, \chi) \neq 0.$$

Consider the expression

$$P(\sigma) = \prod_{\chi \pmod{k}} L(\sigma, \chi).$$

We find that for $\sigma > 1$,

$$\begin{aligned} \ln P(\sigma) &= \sum_{\chi \pmod{k}} \ln L(\sigma, \chi) \\ &= \sum_{\chi \pmod{k}} \sum_{p} \sum_{m \geq 1} \frac{\chi(p^m)}{mp^{m\sigma}} \\ &= \sum_{p} \sum_{m \geq 1} \frac{1}{mp^{m\sigma}} \sum_{\chi \pmod{k}} \chi(p^m)\overline{\chi(1)} \\ &= \sum_{p} \sum_{\substack{m \geq 1 \\ p^m \equiv 1 \pmod{k}}} \frac{1}{mp^{m\sigma}} \geq 0. \end{aligned}$$

Hence, for $\sigma > 1$,

$$P(\sigma) \geq 1. \tag{7.4}$$

Suppose that $L(1,\chi) = 0$ for some χ. Then $L(1,\bar{\chi}) = 0$. Hence, $P(s)$ has two zeros at $s = 1$. But $L(s,\chi_0)$ has a simple pole at $s = 1$, which means that $P(1) = 0$. This is a contradiction to (7.4).

Step 5.

In this final step, we show that for real character $\chi \neq \chi_0$, $L(1,\chi) \neq 0$. Consider the function $f = \chi * u$. Then f is multiplicative by Theorem 2.11. Note that

$$\sum_{l=0}^{m} \chi(p^l) = \begin{cases} 1 & \text{if } p|k \\ \geq 1 & \text{if } p \nmid k, m \text{ even} \\ \geq 0 & \text{if } p \nmid k, m \text{ odd}. \end{cases}$$

Thus, $f(n) \geq 0$ for all n and $f(n) \geq 1$ when n is a square. Hence,

$$F(\sigma) = \sum_{n\geq 1} \frac{f(n)}{n^\sigma} \geq \sum_{n\geq 1} \frac{1}{n^{2\sigma}} = \zeta(2\sigma).$$

In particular, $F(\sigma)$ diverges at $\sigma = 1/2$ and so $\sigma_c \geq 1/2$. By Theorem 6.6, $F(s)$ must have a singularity at $s = \sigma_c \geq 1/2$.

On the other hand, for $\sigma > 1$,

$$F(s) = L(s,\chi)\zeta(s).$$

If $L(1,\chi) = 0$, then $F(s)$ would be analytic in $\sigma > 0$ and hence at $\sigma = \sigma_c$. This contradicts our previous observation that $F(s)$ has a singularity at σ_c and we must have $L(1,\chi) \neq 0$.

From Steps 3 and 4, we conclude that $L(1,\chi) \neq 0$ for all non-principal characters χ. This completes the proof of Dirichlet's Theorem.

Remark 7.2. If p is a prime that satisfies the property that $p+2$ is also a prime, then we call p a twin prime. The *twin primes conjecture* states that there are infinitely many twin primes. This statement remains an open problem.

Motivated by Merten's estimates and the proof of Dirichlet's Theorem of primes in arithmetical progression, it is natural to consider the sum

$$\sum_{p\leq x, p\in T} \frac{1}{p}$$

where T is the set of twin primes. If one can prove that the sum is divergent, then there would be infinitely many primes. Unfortunately, this sum turns out to be convergent (using sieve method). Consequently, this line of attack fails to provide a proof of the twin primes conjecture.

7.6 Exercises

1. Show that any arithmetical function f that is periodic (mod k) and satisfies $f(n) = 0$ if $(n, k) > 1$ can be expressed as a linear combination of characters (mod k).

2. Let $k > 1$ be an integer. Let χ be any non-principal character (mod k). Prove that for all positive integers $a < b$ we have
$$\left|\sum_{n=a}^{b} \chi(n)\right| \leq \frac{1}{2}\varphi(k).$$

3. Construct an infinite set S of primes with the following property: If $p \in S$ and $q \in S$ then
$$\left(\frac{p-1}{2}, \frac{q-1}{2}\right) = (p, q-1) = (p-1, q) = 1.$$

4. Let $c_q(n)$ be given by Exercises 2.6, Problem 3 and define
$$< f, g > \;=\; \lim_{x\to\infty} \frac{1}{x} \sum_{n\leq x} f(n)\overline{g(n)}.$$
Using Section 2.6, Problem 3, show that
$$< c_q, c_{q'} > \;=\; \begin{cases} \varphi(q) & \text{if } q = q' \\ 0 & \text{otherwise.} \end{cases}$$

5. (a) Let $f(x)$ be a polynomial of degree $n \geq 1$ with integer coefficients and suppose for each prime p, an integer m and a prime q exist so that $f(p) = q^m$. Show that for any integer t,
$$q^{m+1} | (f(p + tq^{m+1}) - f(p)).$$

   (b) Deduce from (a) that if $p \neq q$ and $p + tq^{m+1}$ is a prime, then
$$f(p + tq^{m+1}) = q^m.$$

   (c) Derive a contradiction from (b) using Dirichlet's Theorem on primes in arithmetic progression and conclude that $p = q$, or equivalently, that for each prime p, there exists an integer m_p such that
$$f(p) = p^{m_p}.$$

6. (a) Let m and k be positive integers and f be a Dirichlet character (mod k). We say that m is a period of f if $f(n+m) = f(n)$ for all positive integers n. If k is square free, prove that k is the smallest positive period of f.

   (b) Give an example of a Dirichlet character (mod k) for which k is not the smallest positive period of f.

Bibliography

[1] Alfhors, L., *Complex Analysis*, 3rd edn., McGraw-Hill, New York, 1996.

[2] Apostol, T., *Introduction to Analytic Number Theory*, 3rd edn., Springer-Verlag, New York, 1986.

[3] Erdös, P., Ramanujan and I, in *Number Theory, Madras 1987, Proceedings of the International Ramanujan Centenary Conference held at Anna University, Madras, India, Dec.21, 1987, edited by K. Alladi, Lecture Notes in Mathematics (1395)*, Springer-Verlag, Berlin, 1989.

[4] Hildebrand, A., The Prime Number Theorem via the Large Sieve, *Mathematika* **33** (1986) 23–30.

[5] Hildebrand, A., Notes on Analytic Number Theory, University of Illinois at Urbana-Champaign, 1991.

[6] Huxley, H., Exponential sums of lattice points iii, *Proc. London Math. Soc.* **87** (2003) 591–609.

[7] Iwaniec, H. and Mozzochi, C., On the divisor and circle problems, *J. Number Theory* **29**, 1 (1988) 60–93.

[8] Lang, S., *Algebra*, 3rd edn., Addison Wesley, 1993.

[9] Ramanujan, S., A proof of Bertrand's postulate, *J. Indian Math. Soc.* **11** (1919) 181–182.

[10] Rotman, J., *A First Course in Abstract Algebra with Applications*, 3rd edn., Pearson Prentice Hall, 2006.

[11] Tenenbaum, G., *Introduction to Analytic and Probabilistic Number Theory*, Cambridge University Press, 1995.

[12] van der Corput, J., Zum teilerproblem, *Math. Ann.* **98**, 1 (1928) 697–716.

[13] Voronoi, G., Sur une problème du calcul des fonctions asymptotique, *J. Reine Angew. Math.* **126** (1903) 241–282.

[14] Whittaker, E. and Watson, G., *A Course of Modern Analysis*, 4th edn., Cambridge University Press, 1992.

Notation

$a\|b$	a divides b
(a,b)	The greatest common divisor of two integers a and b
$[a,b]$	The lowest common multiple of two integers a and b
$\mu(n)$	The Möbius function
$\sum_{d\|n} f(d)$	sum of values of f over all divisors of n
$[x]$	The integer part of x
$I(n)$	The function is defined by $I(n) = \left[\frac{1}{n}\right]$, $n \in \mathbf{N}$
$\varphi(n)$	The Euler tortient function
$f * g$	The Dirichlet product of two arithmetical functions f and g
$f^{-1}(n)$	The Dirchlet inverse of $f(n)$
$N(n)$	The function is defined by $N(n) = n$, $n \in \mathbf{N}$
$u(n)$	The function is defined by $u(n) = 1$, $n \in \mathbf{N}$
$\Lambda(n)$	The von Mangoldt's function
$c_q(n)$	The Ramanujan sum
$\sigma_\alpha(n)$	The sum of α-th power of all divisors of n
$\lambda(n)$	The Liouville function
$d(n)$	The divisor function
$f(x) = O(g(x))$	This means $f(x) \leq M\|g(x)\|$ for some $M > 0$
$f(x) \ll g(x)$	$f(x) = O(g(x))$
$\sum_{n \leq x} f(n)$	The sum of $f(n)$ over all integers less than or equal to x
$\overline{f}(N)$	$\frac{1}{N} \sum_{n \leq N} f(n)$
$f \sim g$	The quotient $f(x)/g(x)$ approaches 1 as x approaches ∞
$\zeta(s)$	The Riemann zeta function

$\omega(n)$	The number of distinct prime factors of n
$\pi(x)$	The number of primes less than or equal to x
$\psi(n)$	The sum of $\Lambda(k)$ over integers k less than or equal to n
$\sum_{p\leq x} f(p)$	The sum of $f(p)$ over all primes less than or equal to x
$\theta(n)$	The sum of $\ln p$ over primes less than or equal to x
$M(f)$	The value of $\overline{f}(n)$ as n tends to ∞
$\Gamma(z)$	The Gamma function
$L(s,\chi)$	The Dirichlet L-series associated with the character χ

Index